# Cuisenaire Rods® Alphabet Book

by
Maria R. Marolda

Graphics by Diane Phillips

**For Matthew and Ria**

. . . with special thanks to Patricia Moyland and
the staff at Cuisenaire Company for their
helpful suggestions.

Written by: Maria R. Mardola
Cover designed by: Holly Miller
Graphics by: Diane Phillips

© Learning Resources, Inc., Vernon Hills, IL (U.S.A.)
   Learning Resources Ltd., King's Lynn, Norfolk (U.K.)

All rights reserved. This book is copyrighted. No part of this book may be reproduced, stored in a retrieval system, or transmitted, in any form or by any means electronic, mechanical, photocopying, recording, or otherwise, without written permission, except for the specific permission stated below.

Each blackline master or activity page is intended for reproduction in quantities sufficient for classroom use. Permission is granted to the purchaser to reproduce each blackline master or activity page in quantities suitable for noncommercial classroom use.

The word Cuisenaire® and the color sequence of the rods, cubes, and squares are registered trademarks of ETA/Cuisenaire.

ISBN  1-56911-061-1

Printed in China.

# INTRODUCTION

What you see when you open the **Cuisenaire® Rods Alphabet Book** is a simple, two-page format for each letter, A through Z. The left-hand page has two versions of the letter, one with open space, the other gridded into centimeter squares, together with some problems to be solved with Cuisenaire® Rods. The companion right-hand page contains pictures of objects which begin with the letter and are to be covered with the rods. What is not obvious upon first glance is the mathematical richness and depth these experiences provide.

The book is used as early as age 5 (Kindergarten) and as late as it still seems fun (Adults, Try Some!!). The purpose and focus can be modified to fit the age and needs of the users. Young children may use the book to learn the rod colors, to count, to become more acquainted with the letters of the alphabet, and to develop space-filling ability. The older mind will seize upon the opportunities for problem-solving challenges similar to Tangram Puzzles, but which lend themselves naturally to the formulation of addition sentences and the use of multiplication facts. The motivating, self-checking nature of these exercises makes them extremely easy to use for children of all ages.

No prior experience with rods is needed to begin. The questions on the left-hand pages have been kept basically the same for each letter of the alphabet so that children can work individually or in small groups without much help. Similar questions should be posed and discussed for the life-objects on the right-hand pages. Doing a few sheets together can help the beginning reader establish a pattern for proceeding through the other pages.

Since it is helpful for all children to call the rods by the same color names, the standard ones are listed below. Colors may be recorded using crayons to match the rod colors, using color words, or by using the short-hand codes for the colors (first letters for seven of the colors and last letters for the three "b" color names: blacK, browN, bluE). Establishing the numerical value of the rods in terms of white rods gives children practice with whole number addition and multiplication concepts. As they copy the rod pictures and letters of the alphabet using different collections of rods, children find that more than one number sentence for the same sum can be generated. The aspect of more than one right answer for the same problem enhances the fun and excitement of the book.

## CUISENAIRE ROD CHART

| Color Names | Rod Codes | Numerical Value When White = 1 |
|---|---|---|
| White | W | 1 |
| Red | R | 2 |
| Green | G | 3 |
| Purple | P | 4 |
| Yellow | Y | 5 |
| Dark green | D | 6 |
| blacK | K | 7 |
| browN | N | 8 |
| bluE | E | 9 |
| Orange | O | 10 |

# MATHEMATICAL CONCEPTS

The **Cuisenaire® Rods Alphabet Book** provides concrete experiences with the following mathematical concepts. The use of Cuisenaire® Rods places these mathematical ideas into an exciting problem solving context.

**Counting:** The activities provide two types of counting exercises — counting the number of colors used and the number of rods used. Whenever more than one rod of the same color is used, the number of rods will be greater than the number of colors used. The problems which involve covering the alphabet letters with white rods require children to count as high as 34.

**One-to-One Correspondence:** Matching white rods to the centimeter squares gives concrete experience with the concept of one-to-one correspondence. Also, placing the 3-dimensional rods on the 2-dimensional representation helps children develop spatial awareness, perceptual ability, and fine-motor coordination.

**Spatial Relationships:** Placing the rods on the pictures so they fit exactly within the given outlines is a challenging task and a growing experience for very young or perceptually handicapped children who might tend to go outside the borders. Coloring the rod lengths on the centimeter grids provides for all children the necessary transitions from the rod lengths to the numerical value in terms of white rods. When rods are measured in terms of white rods, whole number relationships evolve. (It should be noted that the white rod does not always have to be considered as 1. However, using other rods as 1 would lead to fraction work well beyond the intended scope of this book.)

**Symmetry:** Many arrangements of rods are symmetric horizontally or vertically. It is an interesting exercise to find all symmetric letters and to color them so that the colors of rods used also show symmetry.

**Addition Sentences:** When rods are used to cover the gridded letters, the numerical values of the rods can be added.
> For example, when the A is covered with a Red, blacK, White, browN, and Green rod, the rod story for the letter A becomes: R + K + W + N + G. The numerical sentence, when white is 1, becomes: 2 + 7 + 1 + 8 + 3 = 21. This sum equals the total number of white rods needed to cover the A. It also equals the total number of centimeter squares on the gridded letter A.

**More Than One Right Answer:** Since the alphabet letters and rod pictures can be covered with more than one collection of rods, there are many right answers to most questions.
> For example, the letter A can also be covered by a Purple, Yellow, Green, White, Dark green and Red rod, giving the rod story P + Y + G + W + D + R and the numerical sentence 4 + 5 + 3 + 1 + 6 + 2 = 21.

In the section **Answers and Suggestions** at the end of the book, sample rod stories and numerical sentences are given for each letter of the alphabet.

**Multiplication Facts:** When children cover an alphabet letter with rods of all the same color, the repeated addition sentence can be thought of as multiplication.
> For example, the letter A can be covered with 7 light green rods. The addition sentence 3 + 3 + 3 + 3 + 3 + 3 + 3 = 21 can be thought of as 7 × 3 = 21.

Children can also shorten some of their rod stories using multiplication within their numerical sentence.

> For example, the letter B can be covered with P + P + G + W + W + R + R + R + R + R + R + R. The numerical sentence can be thought of as (2 × 4) + (1 × 3) + (2 × 1) + (7 × 2) = 27.

It is also interesting for children to see that some letters, like the X, cannot be covered with rods of all the same color (other than white rods). Even though the letter X has a value of 21 whites, it cannot be covered with 7 light green rods nor 3 black rods. A challenging problem solving task would be to find all the alphabet letters and rod pictures that can be covered with rods of all the same color (other than white rods which always work).

**Area:** Even though the concept of area is not formalized in this book, children are finding the number of square units in each of the alphabet letters. For example, since the letter A contains 21 centimeter squares, its area is 21 square centimeters. For older or more gifted students, the concept of perimeter could be introduced. What is the distance in centimeters around each letter? Do the letters with equal areas necessarily have equal perimeters?

**Mathematics Vocabulary:** Children actively use many mathematical words in doing these puzzles. They may be asked to use <u>exactly</u> 9 rods, <u>more than 8</u> rods, <u>less than 6</u> rods, <u>the least number</u> of rods, or <u>only</u> 2 colors of rods. Children find that working concretely with rods gives these words meaning. Once the vocabulary is established, they should be encouraged to discuss their own work with the rod pictures in precise mathematical terms. The mathematics vocabulary words and resulting concepts being developed for each letter are delineated in the section **Answers and Suggestions**.

**Problem Solving:** The entire book is presented in a problem solving spirit. However, the greatest problem solving challenge for each alphabet letter is to make it with all different rods (that is, no two rods of the same color). As children start to build the letter, they may have to make adjustments and find substitutions to complete the challenge. Many of the letters can be done in more than one way. The task of using <u>all different</u> rods is fun and motivating for all ages and should be applied to the rod pictures as well.

# ADDITIONAL SUGGESTIONS

1. In covering the rod pictures, the children should not place rods on the screened or blackened portion of the pictures.
2. Children should make their own rod pictures for each alphabet letter. They enjoy having a blank sheet of paper as a defined work space. In the section, **Answers and Suggestions**, four suggestions of further things to make are given for each alphabet letter. Also, a glossary summarizing all alphabet words used in the rod pictures on the right-hand pages may be found on the last page of the book.
3. The order of presentation of the alphabet letters can be coordinated with the language development of the children.
4. By the very nature of the geometric shapes of the alphabet letters, some generate larger numbers and more difficult numerical sentences than others. See the **Answers and Suggestions** for a discussion of the mathematics for each letter.
5. The answers are not complete, as there are seemingly endless ways to do some letters. The given samples focus on the particular questions asked for each letter. There are at least two sample answers to each of the problem solving questions.
6. Even though the book can be used individually without much adult input, it lends itself nicely to math sharing discussions. Children enjoy telling about their different ways of doing the problems. One should capitalize on this opportunity for practice with potentially complex addition sentences which children love to solve.

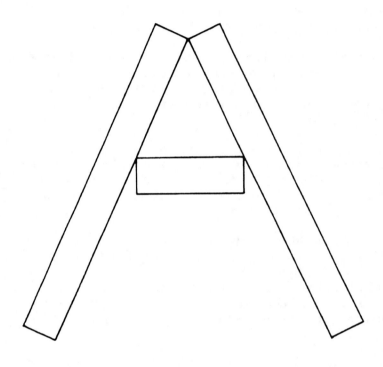

1. Build the A with rods. What colors did you use?

   _____

   How many rods did you use?

   _____

2. Use only light green rods to build the A.

   How many did you use?

   _____

3. Build the A with rods of all different colors.

   Color your A.

4. Cover each color with white rods.

   How many white rods did you need for each color?

   _____

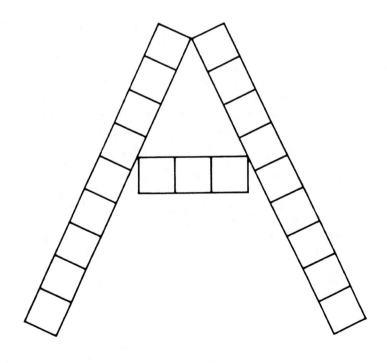

Cuisenaire® Rods Alphabet Book   © Learning Resources, Inc.

# A a is for...

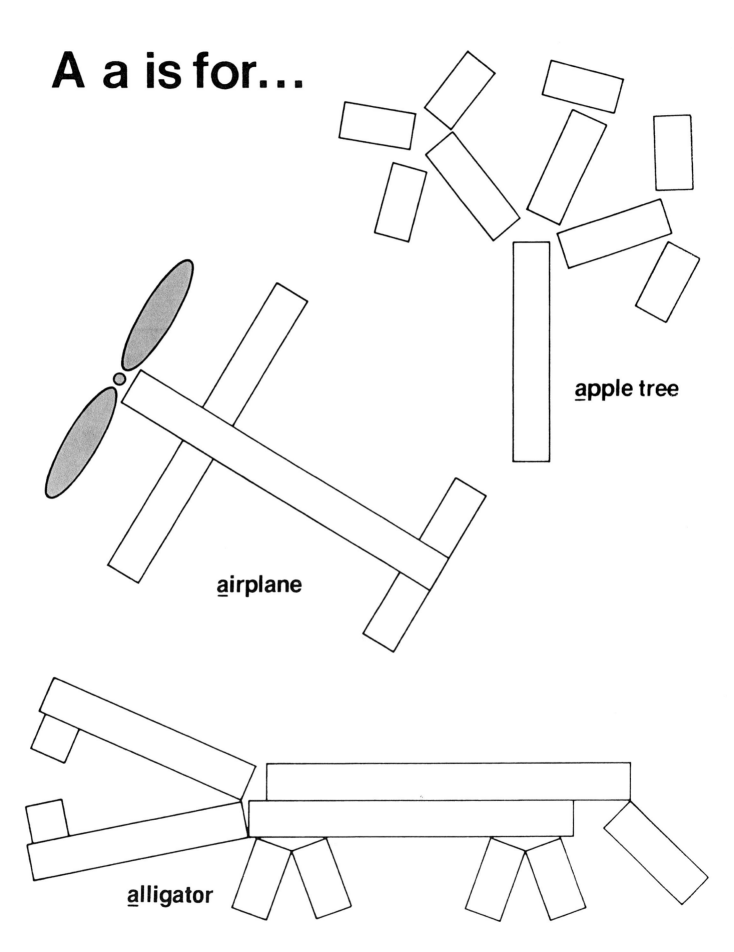

**_a_pple tree**

**_a_irplane**

**_a_lligator**

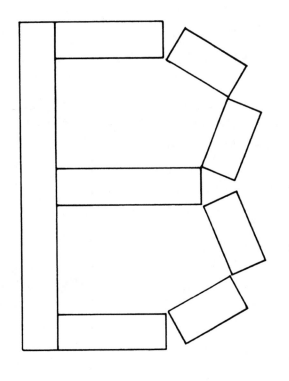

1. Build the B with rods. What rods did you use?

   _____

   How many rods did you use?

   _____

2. Use <u>exactly</u> <u>9</u> rods to build the B.

   What rods did you use?

   _____

3. Build the B using <u>only</u> <u>2</u> colors.

   Color your B.

4. Cover each color with white rods.

   How many white rods did you need for each color?

   _____

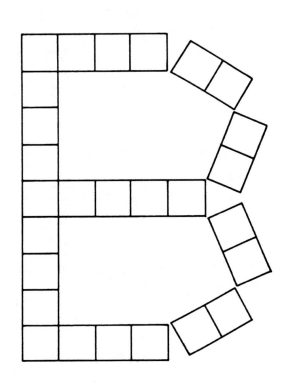

Cuisenaire® Rods Alphabet Book   © Learning Resources, Inc.

# B b is for...

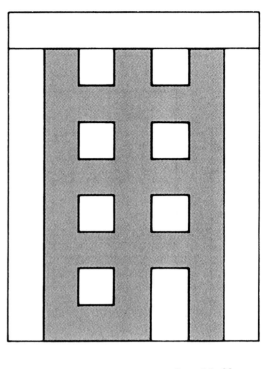
**b**uilding

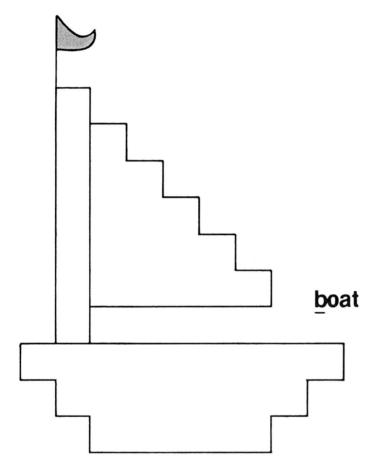
**b**oat

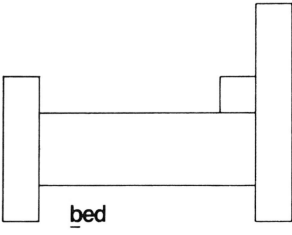
**b**ed

1. Build the C with rods. What colors did you use?

   _____

   How many rods did you use?

   _____

2. Use the least number of rods to build the C.

   What rods did you use?

   _____

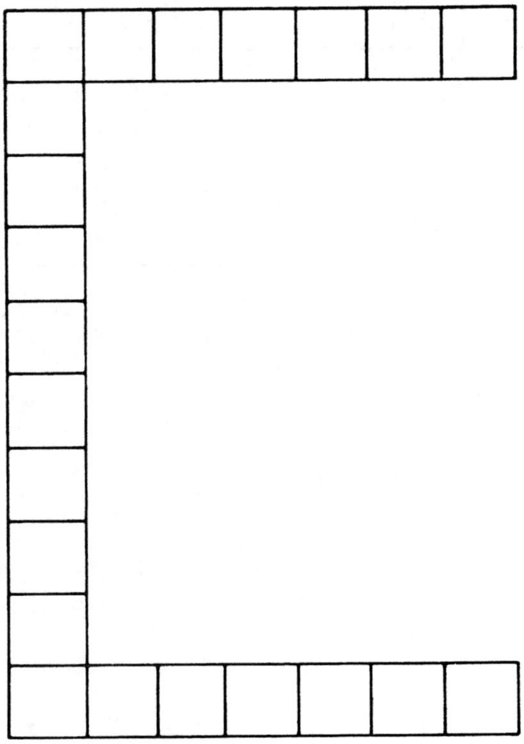

3. Build the C with rods of all different colors.

   Color your C.

4. Cover each color with white rods.

   How many white rods did you need for each color?

   _____

Cuisenaire® Rods Alphabet Book   © Learning Resources, Inc.

# C c is for...

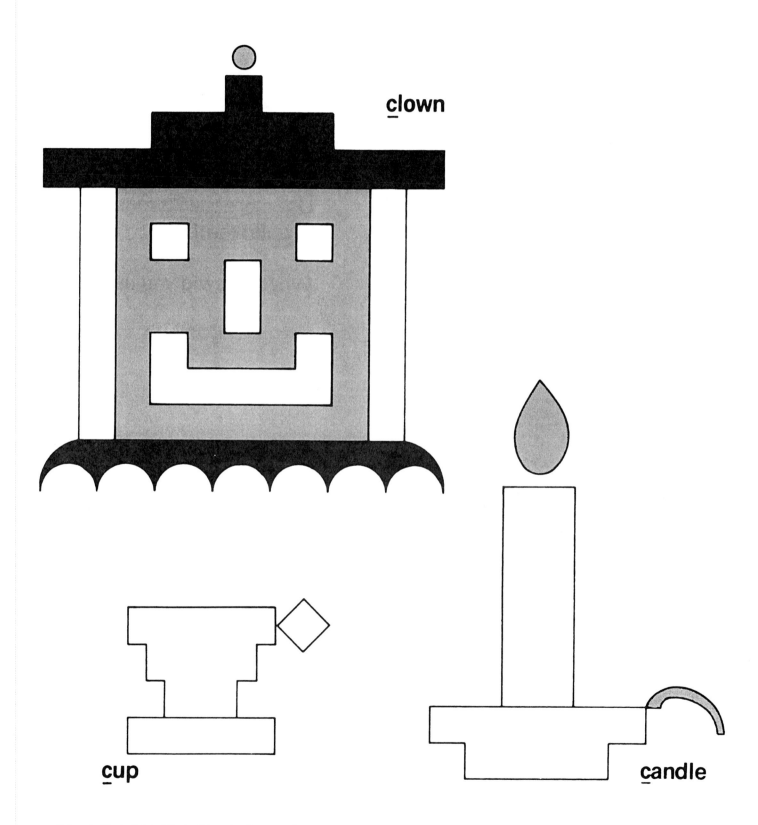

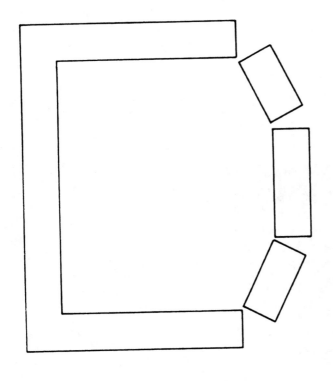

1. Build the D with rods. What colors did you use?

   _____

   How many rods did you use?

   _____

2. Use <u>more than 6</u> rods to build the D.

   What rods did you use?

   _____

3. Build the D using <u>only</u> <u>2</u> colors.

   Color your D.

4. Cover each color with white rods.

   How many white rods did you need for each color?

   _____

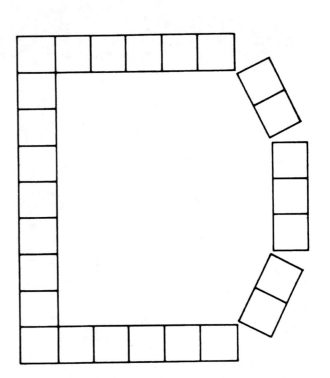

Cuisenaire® Rods Alphabet Book    © Learning Resources, Inc.

# D d is for...

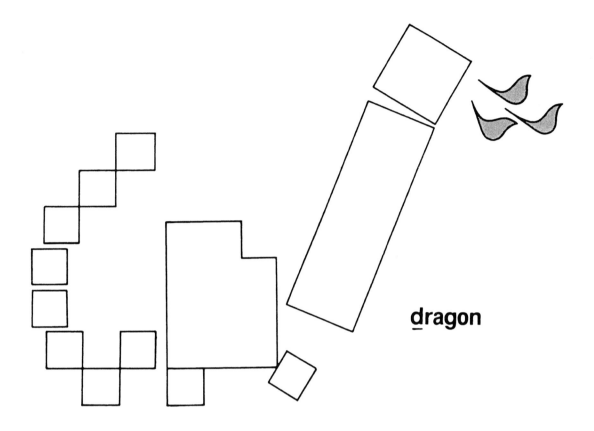

**dragon**

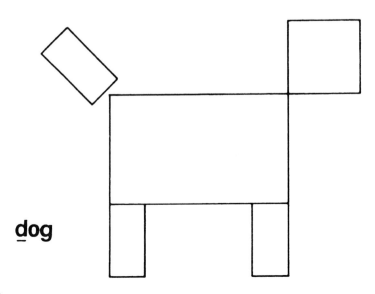

**dog**

1. Build the E with rods. What colors did you use?

   _____

   How many rods did you use?

   _____

2. Use only red rods to build the E.

   How many red rods did you use?

   _____

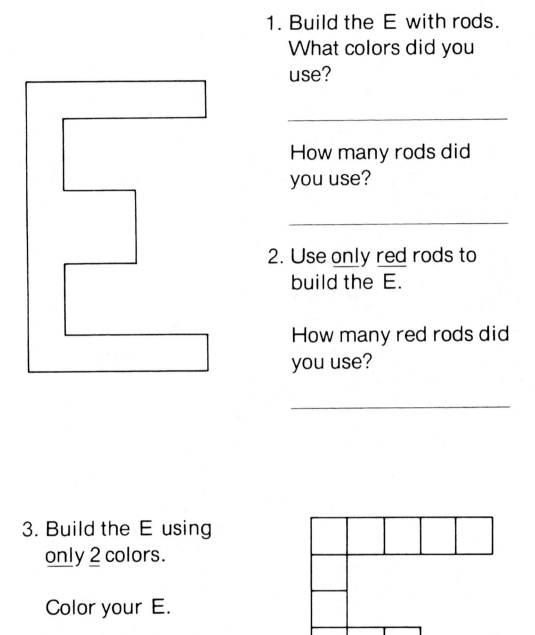

3. Build the E using only 2 colors.

   Color your E.

4. Cover each color with white rods.

   How many white rods did you need for each color?

   _____

Cuisenaire® Rods Alphabet Book    © Learning Resources, Inc.

# E e is for...

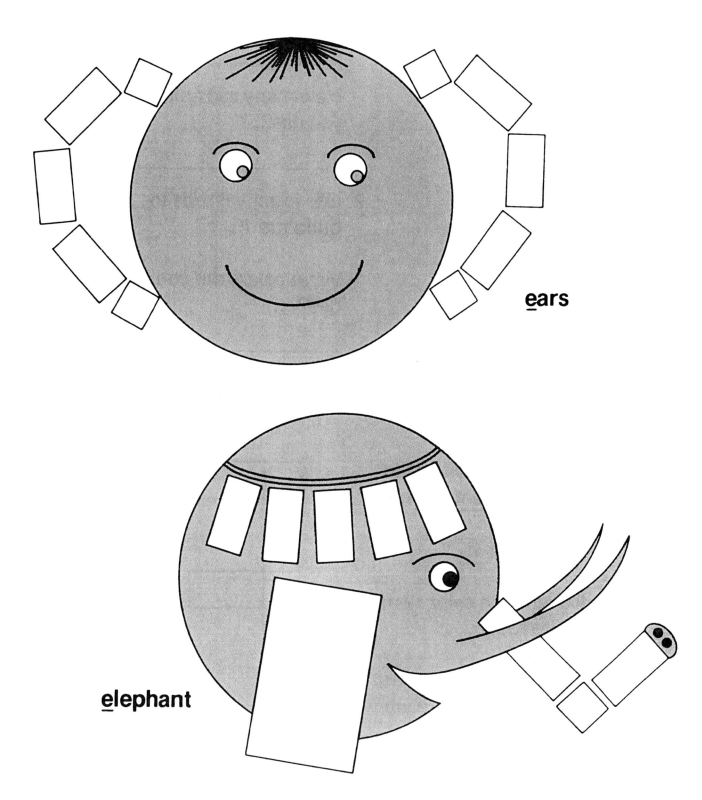

**ears**

**elephant**

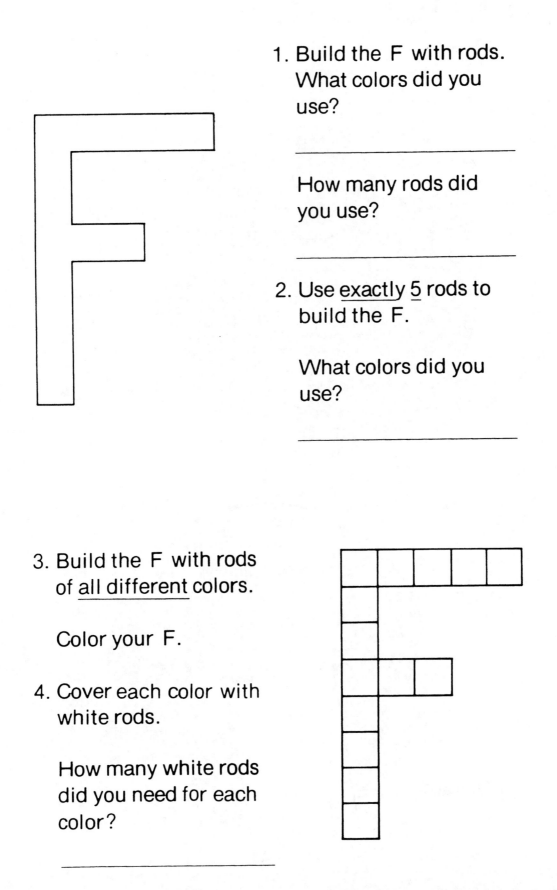

1. Build the F with rods. What colors did you use?

   _____

   How many rods did you use?

   _____

2. Use <u>exactly</u> <u>5</u> rods to build the F.

   What colors did you use?

   _____

3. Build the F with rods of <u>all different</u> colors.

   Color your F.

4. Cover each color with white rods.

   How many white rods did you need for each color?

   _____

Cuisenaire® Rods Alphabet Book   © Learning Resources, Inc

# F f is for...

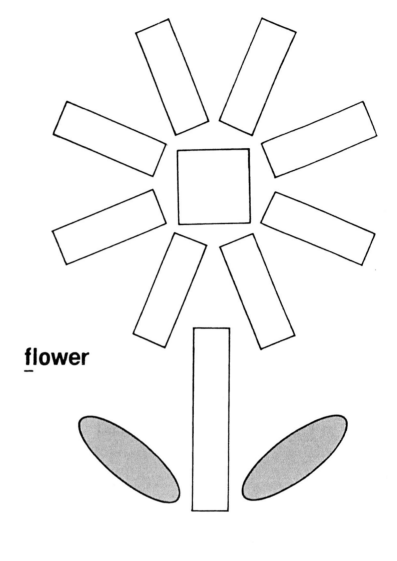
**flower**

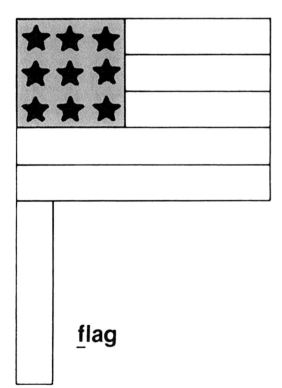
**flag**

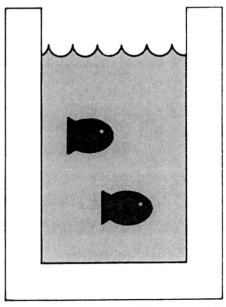
**fish tank**

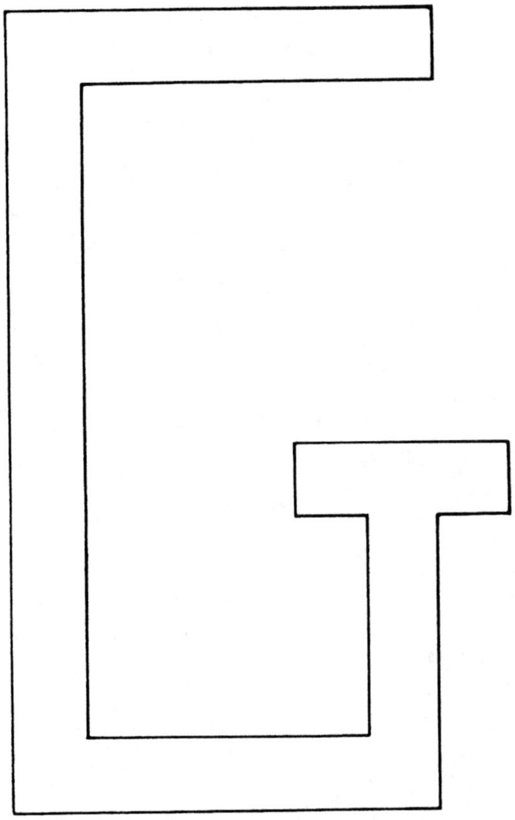

1. Build the G with rods. What colors did you use?

   _____

   How many rods did you use?

   _____

2. Use <u>only</u> <u>light</u> <u>green</u> rods to build the G.

   How many green rods did you use?

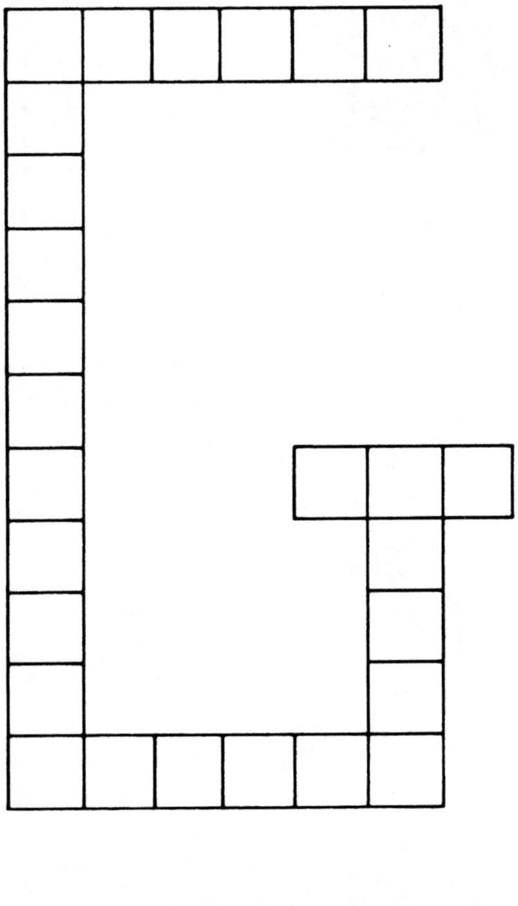

   _____

3. Build the G with rods of <u>all different</u> colors.

   Color your G.

4. Cover each color with white rods.

   How many white rods did you need for each color?

   _____

Cuisenaire® Rods Alphabet Book    © Learning Resources, Inc.

# G g is for...

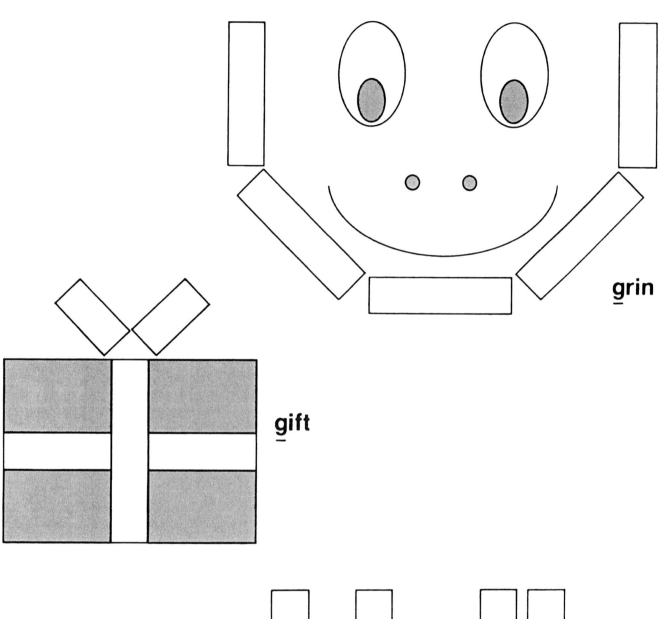

grin

gift

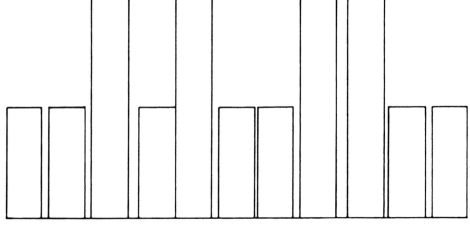

grass

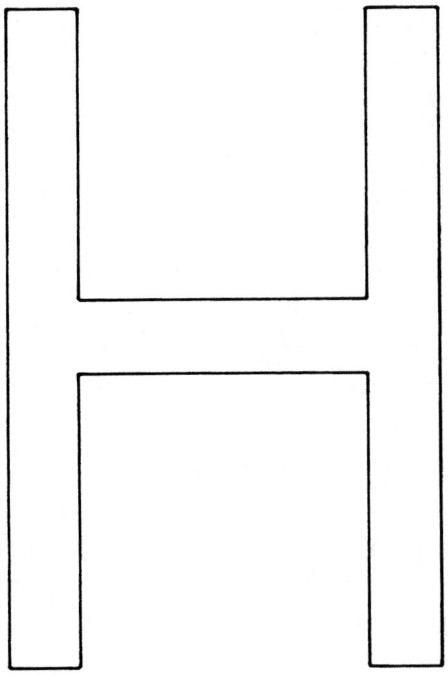

1. Build the H with rods. What colors did you use?

   _____

   How many rods did you use?

   _____

2. Use <u>less than 6</u> rods to build the H.

   What colors did you use?

   _____

3. Build the H with rods of <u>all different</u> colors.

   Color your H.

4. Cover each color with white rods.

   How many white rods did you need for each color?

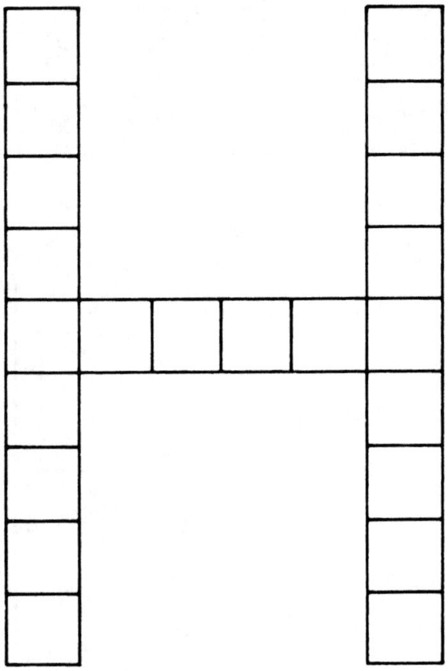

   _____

Cuisenaire® Rods Alphabet Book   © Learning Resources, Inc.

# H h is for...

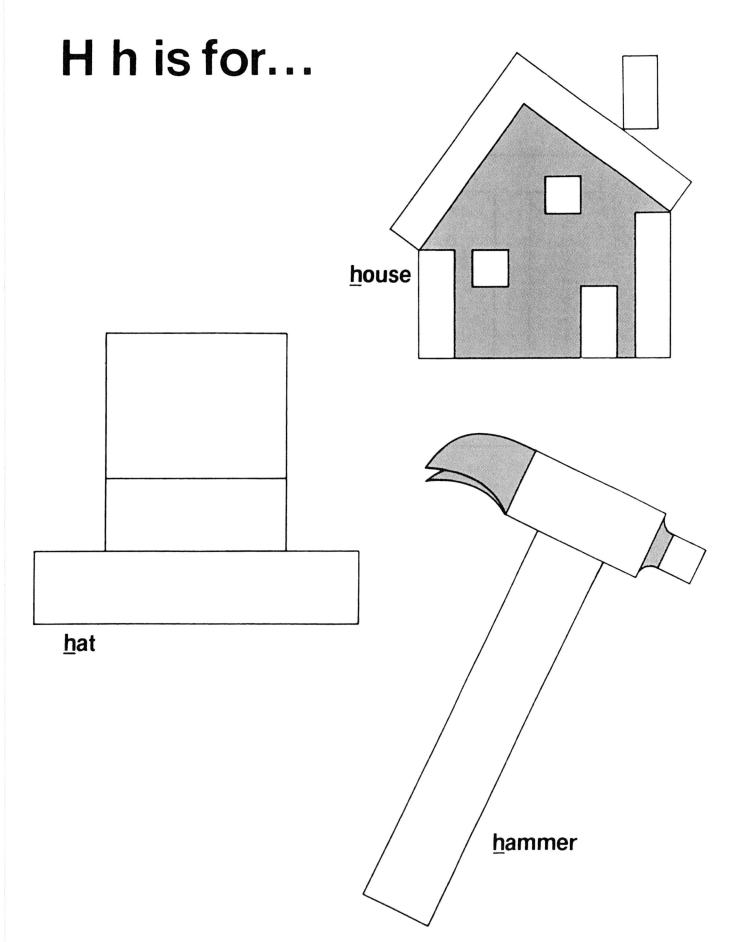

house

hat

hammer

1. Build the I with rods. What colors did you use?

   _____

   How many rods did you use?

   _____

2. Use only one color to build the I.

   What color did you use?

   _____

3. Build the I using rods of all different colors.

   Color your I.

4. Cover each color with white rods.

   How many white rods did you need for each color?

   _____

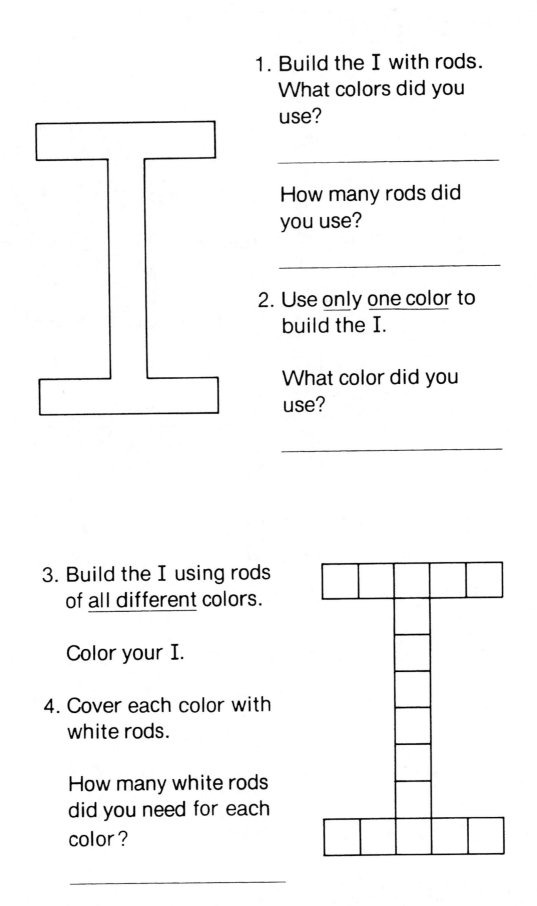

Cuisenaire® Rods Alphabet Book  © Learning Resources, Inc.

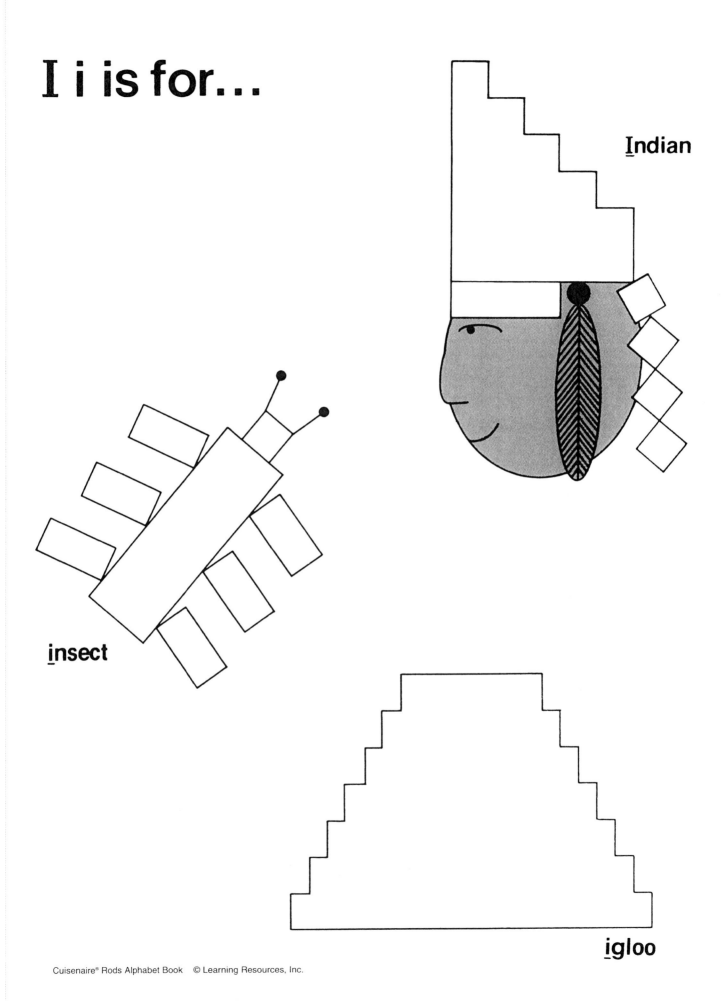

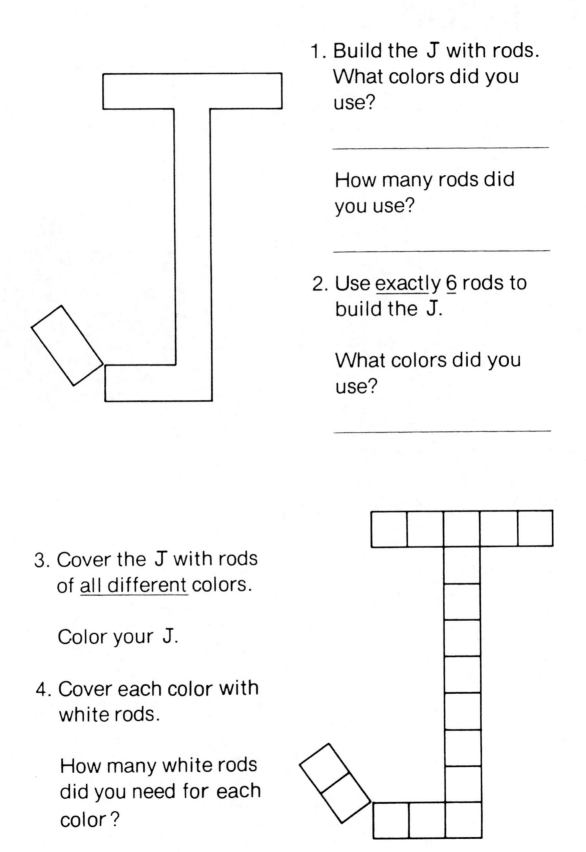

1. Build the J with rods. What colors did you use?

   _____

   How many rods did you use?

   _____

2. Use <u>exactly</u> <u>6</u> rods to build the J.

   What colors did you use?

   _____

3. Cover the J with rods of <u>all different</u> colors.

   Color your J.

4. Cover each color with white rods.

   How many white rods did you need for each color?

   _____

Cuisenaire® Rods Alphabet Book   © Learning Resources, Inc.

# J j is for...

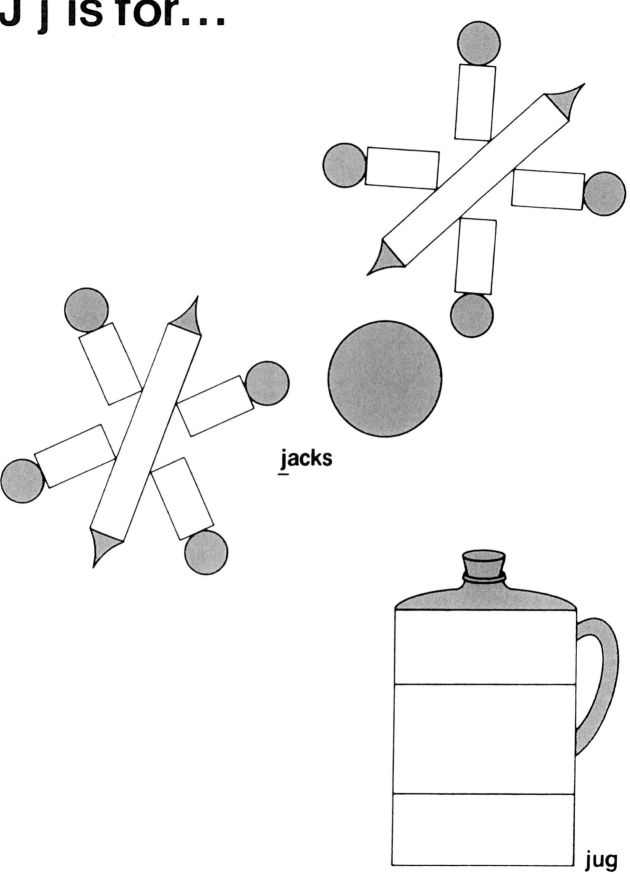

jacks

jug

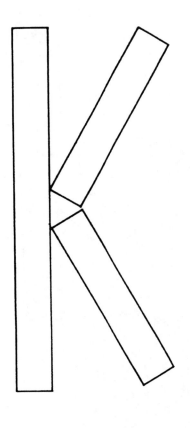

1. Build the K with rods. What colors did you use?

   _____

   How many rods did you use?

   _____

2. Use only yellow rods to build the K.

   How many yellow rods did you use?

   _____

3. Build the K with rods of all different colors.

   Color the K.

4. Cover each color with white rods.

   How many white rods did you need for each color?

   _____

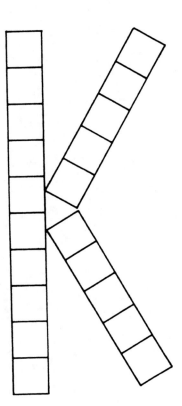

Cuisenaire® Rods Alphabet Book   © Learning Resources, Inc.

# K k is for...

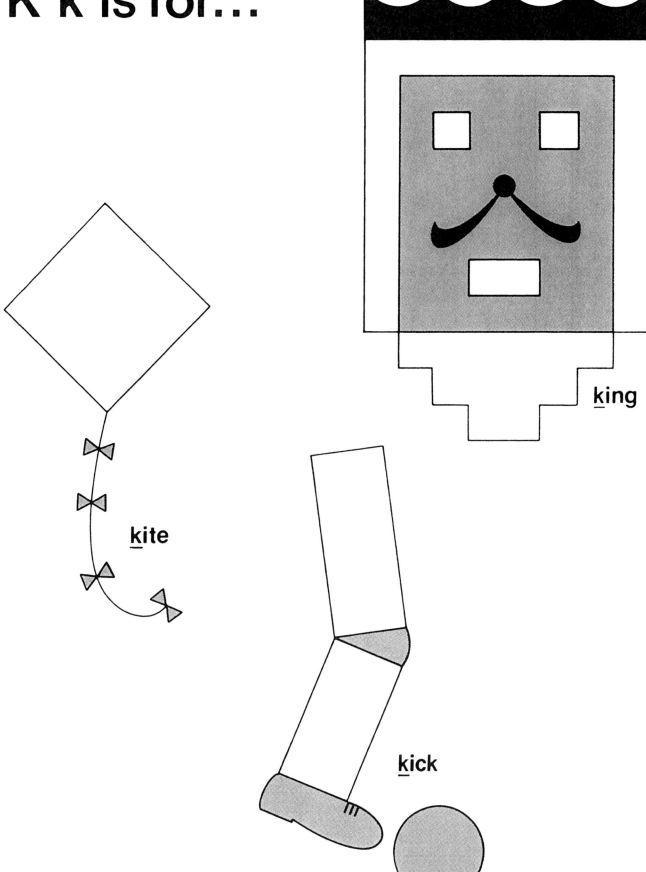

_k_ing

_k_ite

_k_ick

1. Build the L with rods. What colors did you use?

   _____

   How many rods did you use?

   _____

2. Use <u>exactly</u> <u>5</u> rods to build the L.

   What colors did you use?

   _____

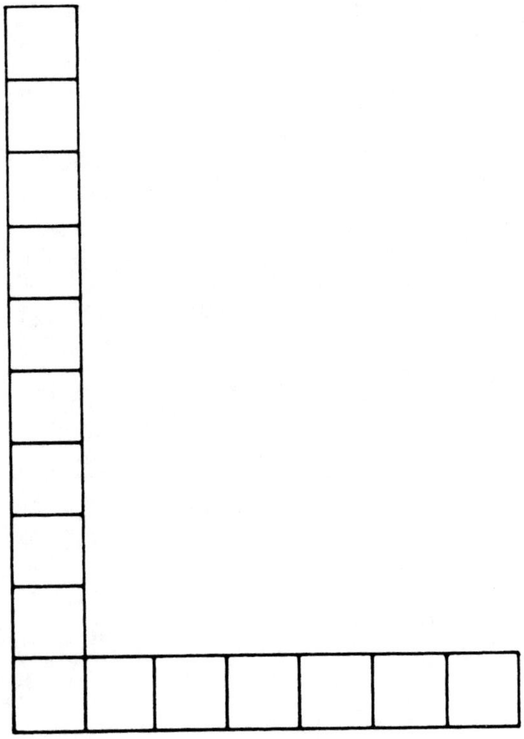

3. Build the L with rods of <u>all different</u> colors.

   Color your L.

4. Cover each color with white rods.

   How many white rods did you need for each color?

   _____

# L l is for...

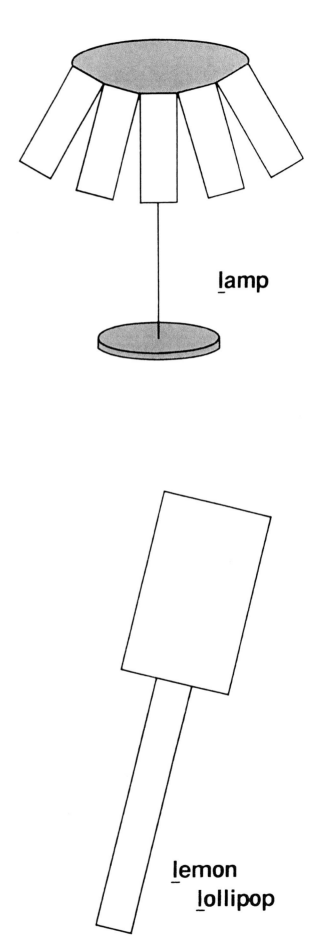

lamp

lemon
lollipop

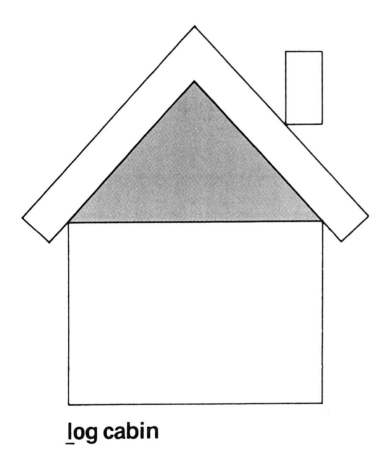

log cabin

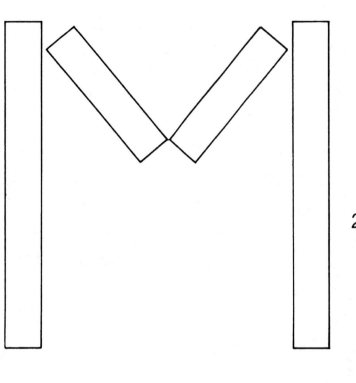

1. Build the M with rods. What colors did you use?

   _____

   How many rods did you use?

   _____

2. Use the least number of rods to build the M.

   What colors did you use?

   _____

3. Build the M with rods of all different colors.

   Color your M.

4. Cover each color with white rods.

   How many white rods did you need for each color?

   _____

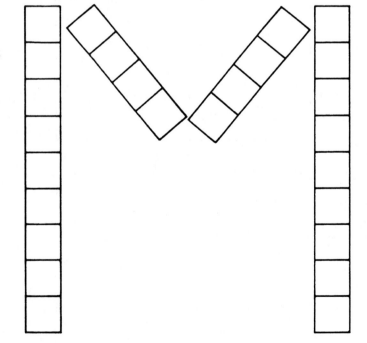

Cuisenaire® Rods Alphabet Book   © Learning Resources, Inc.

# M m is for...

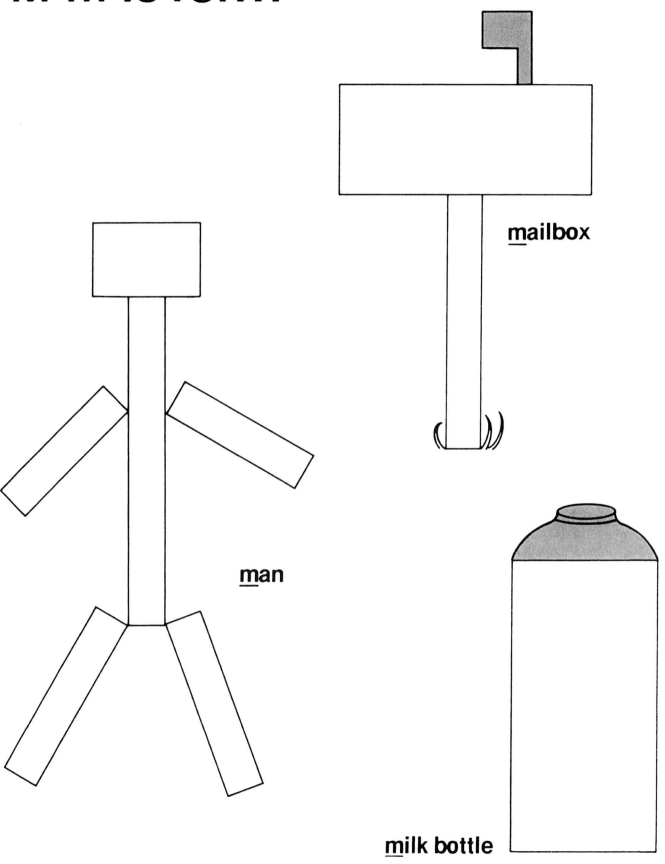

mailbox

man

milk bottle

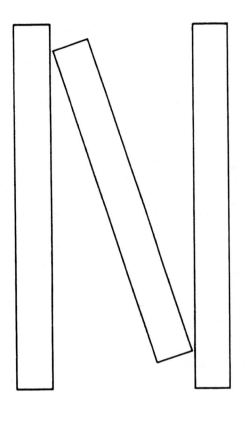

1. Build the N with rods. What colors did you use?

   _____

   How many rods did you use?

   _____

2. Use <u>more than 7</u> rods to build the N.

   What colors did you use?

   _____

3. Build the N with rods of <u>all different</u> colors.

   Color your N.

4. Cover each color with white rods.

   How many white rods did you need for each color?

   _____

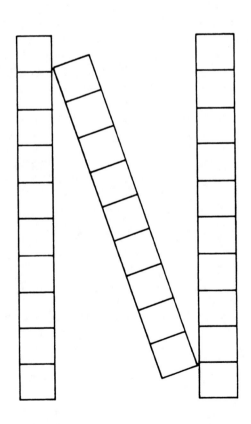

# N n is for...

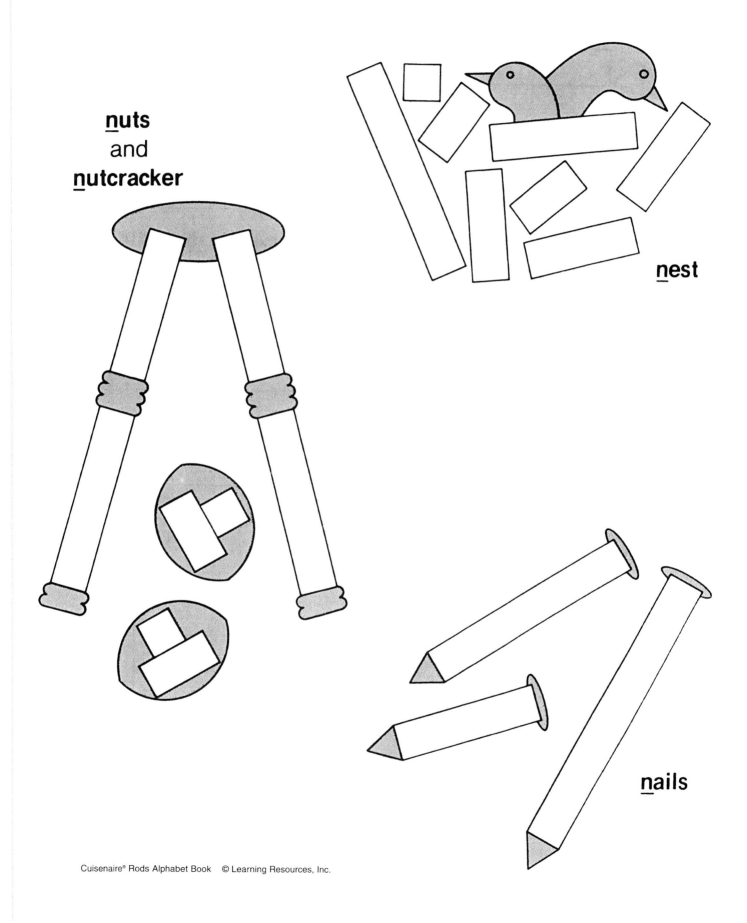

**n**uts
and
**n**utcracker

**n**est

**n**ails

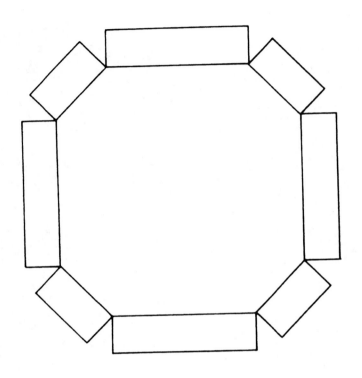

1. Build the O with rods. What colors did you use?

   _____

   How many rods did you use?

   _____

2. Use <u>exactly</u> <u>10</u> rods to build the O.

   What colors did you use?

   _____

3. Build the O using <u>only</u> <u>2</u> colors.

   Color your O.

4. Cover each color with white rods.

   How many white rods did you need for each color?

   _____

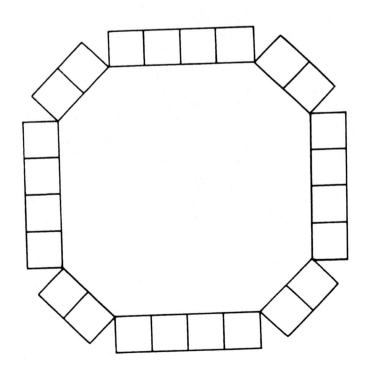

Cuisenaire® Rods Alphabet Book   © Learning Resources, Inc.

# O o is for...

**o**ctopus

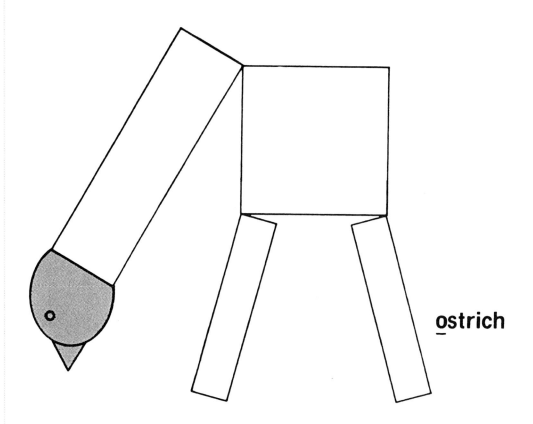

**o**strich

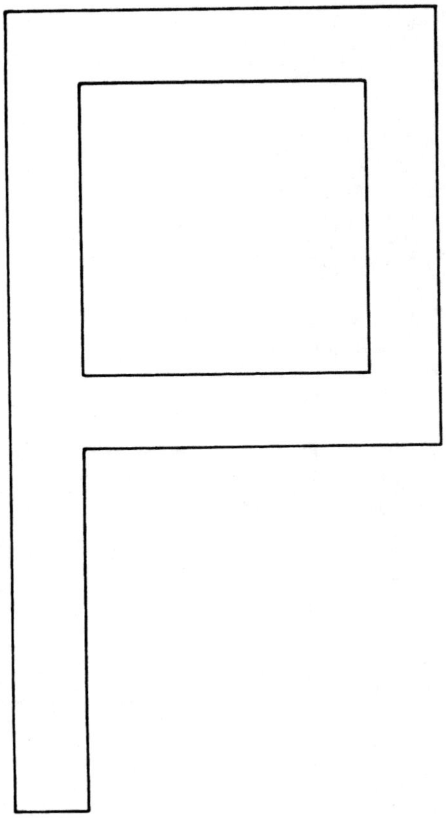

1. Build the P with rods. What colors did you use?

   _____

   How many rods did you use?

   _____

2. Use only <u>3</u> <u>colors</u> to build the P.

   How many yellow rods are needed?

   _____

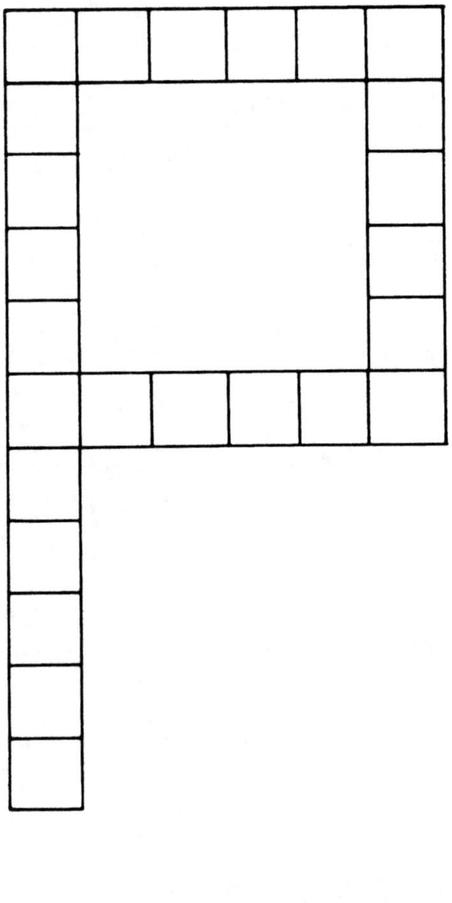

3. Build the P with rods of <u>all different</u> colors.

   Color your P.

4. Cover each color with white rods.

   How many white rods did you need for each color?

   _____

Cuisenaire® Rods Alphabet Book   © Learning Resources, Inc.

# P p is for...

pail
and
paintbrush

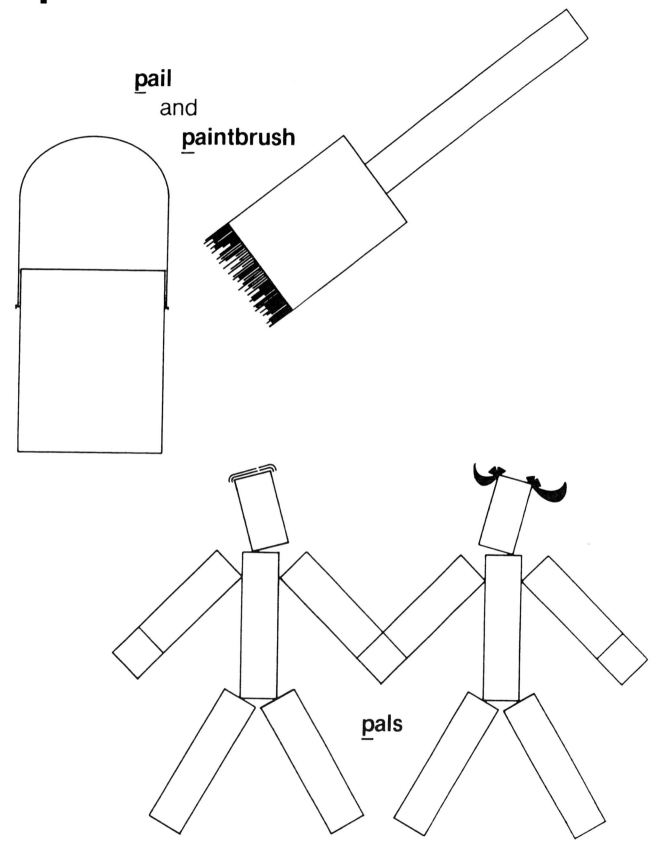

pals

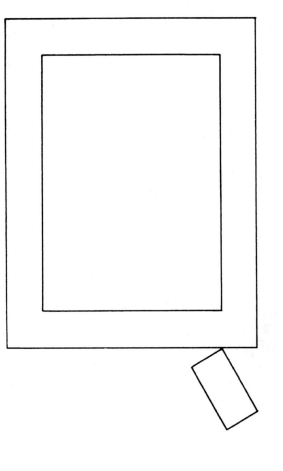

1. Build the Q with rods. What colors did you use?

   _____

   How many rods did you use?

   _____

2. Use the least number of rods to make the Q.

   What colors did you use?

   _____

3. Build the Q with rods of all different colors.

   Color your Q.

4. Cover each color with white rods.

   How many white rods did you need for each color?

   _____

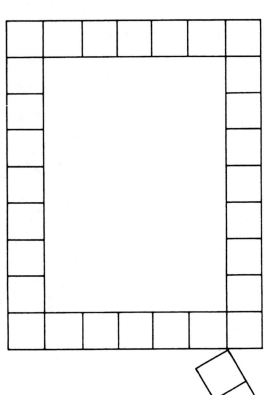

Cuisenaire® Rods Alphabet Book   © Learning Resources, Inc.

# Q q is for...

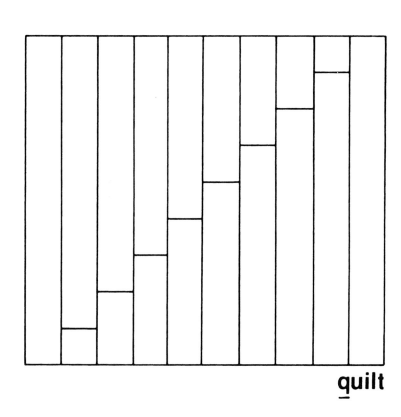

**q**uilt

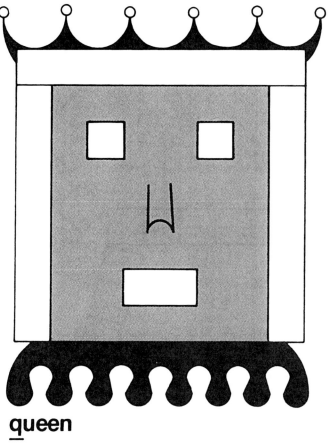

**q**ueen

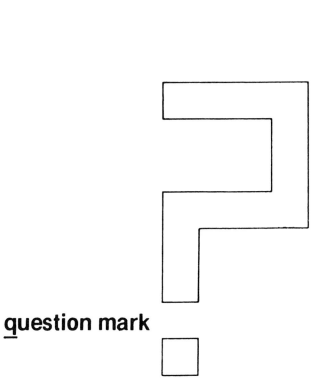

**q**uestion mark

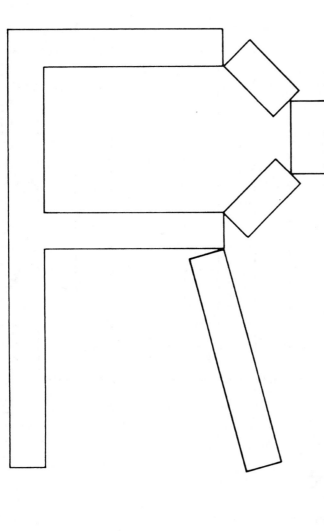

1. Build the R with rods. What colors did you use?

   _____

   How many rods did you use?

   _____

2. Use <u>exactly</u> <u>8</u> rods to build the R.

   How many red rods did you use?

   _____

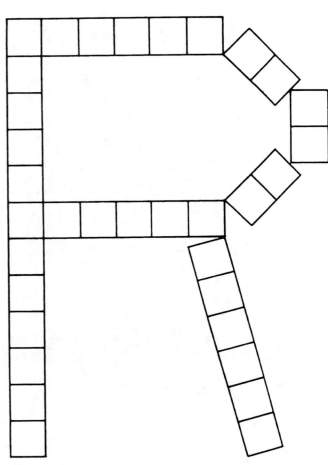

3. Build the R using <u>only</u> <u>2</u> colors.

   Color your R.

4. Cover each color with white rods.

   How many white rods did you use for each color?

   _____

Cuisenaire® Rods Alphabet Book   © Learning Resources, Inc.

# R r is for...

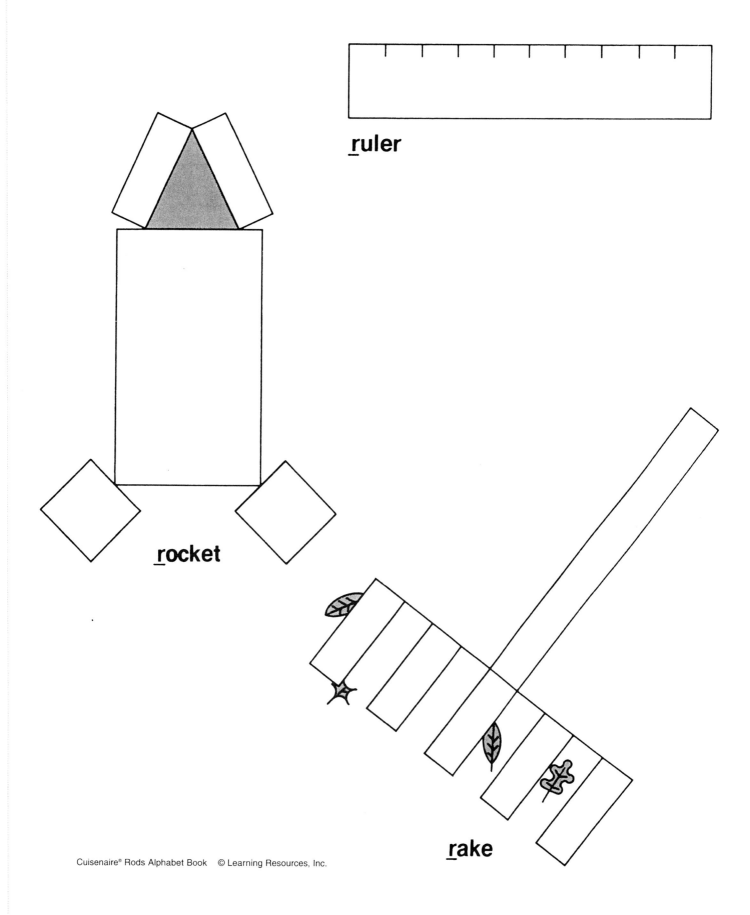

_ruler

_rocket

_rake

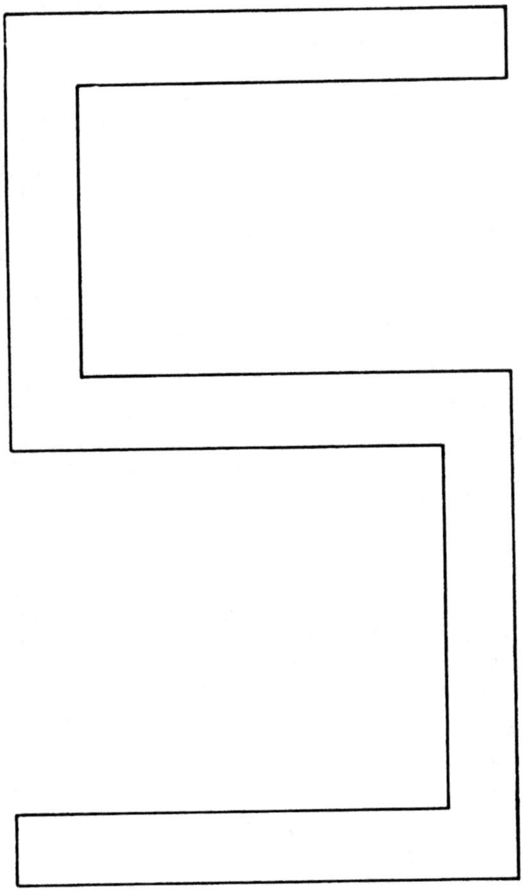

1. Build the S with rods. What colors did you use?

　　_____

　How many rods did you use?

　　_____

2. Use <u>only light green</u> rods to build the S.

　How many green rods did you use?

　　_____

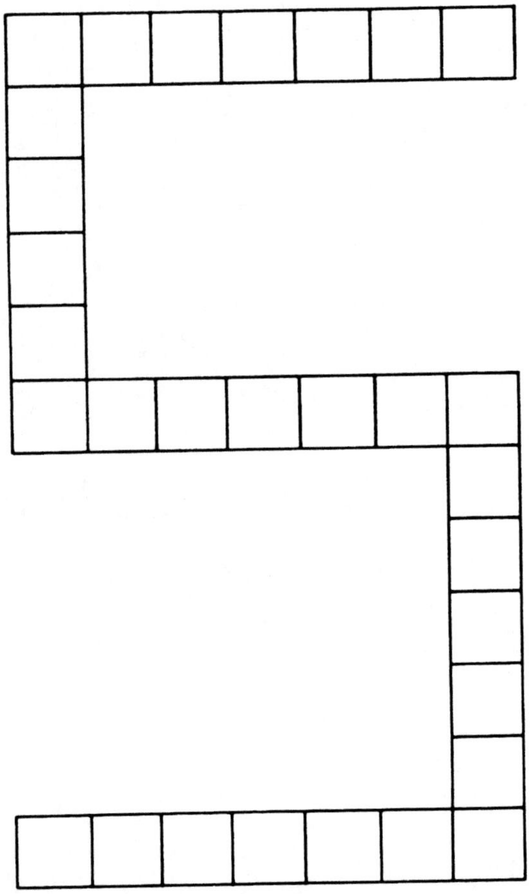

3. Build the S using <u>only 2</u> colors.

　Color your S.

4. Cover each color with white rods.

　How many white rods did you use for each color?

　　_____

Cuisenaire® Rods Alphabet Book　© Learning Resources, Inc.

# S s is for...

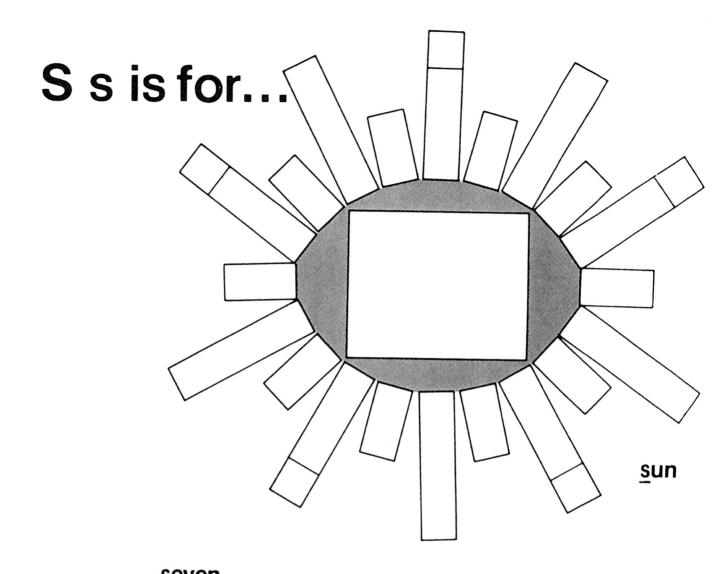

sun

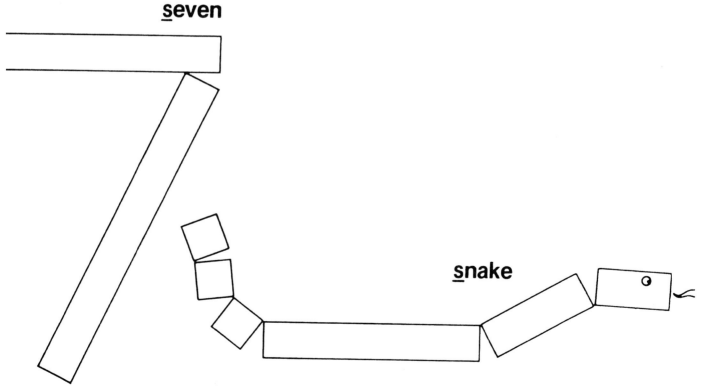

seven

snake

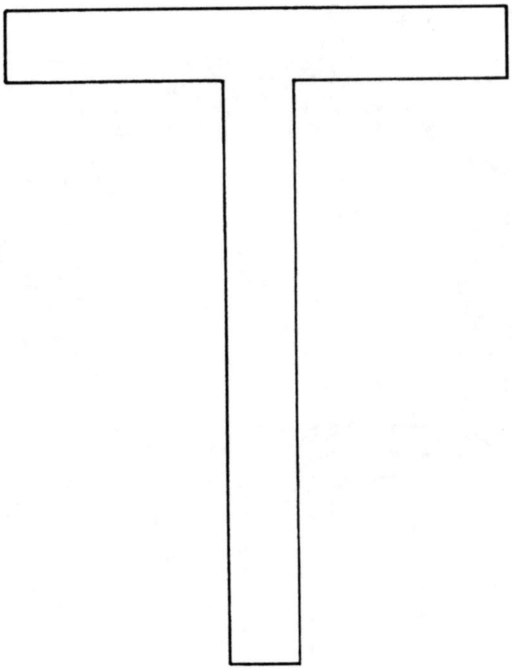

1. Build the T with rods. What colors did you use?

   _____

   How many rods did you use?

   _____

2. Use <u>only one color</u> to build the T.

   What color did you use?

   _____

3. Build the T with rods of <u>all different</u> colors.

   Color your T.

4. Cover each color with white rods.

   How many white rods did you need for each color?

   _____

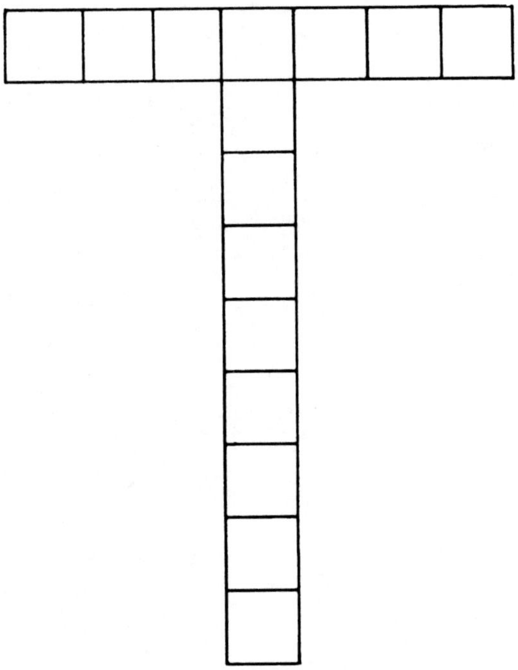

Cuisenaire® Rods Alphabet Book   © Learning Resources, Inc.

# T t is for...

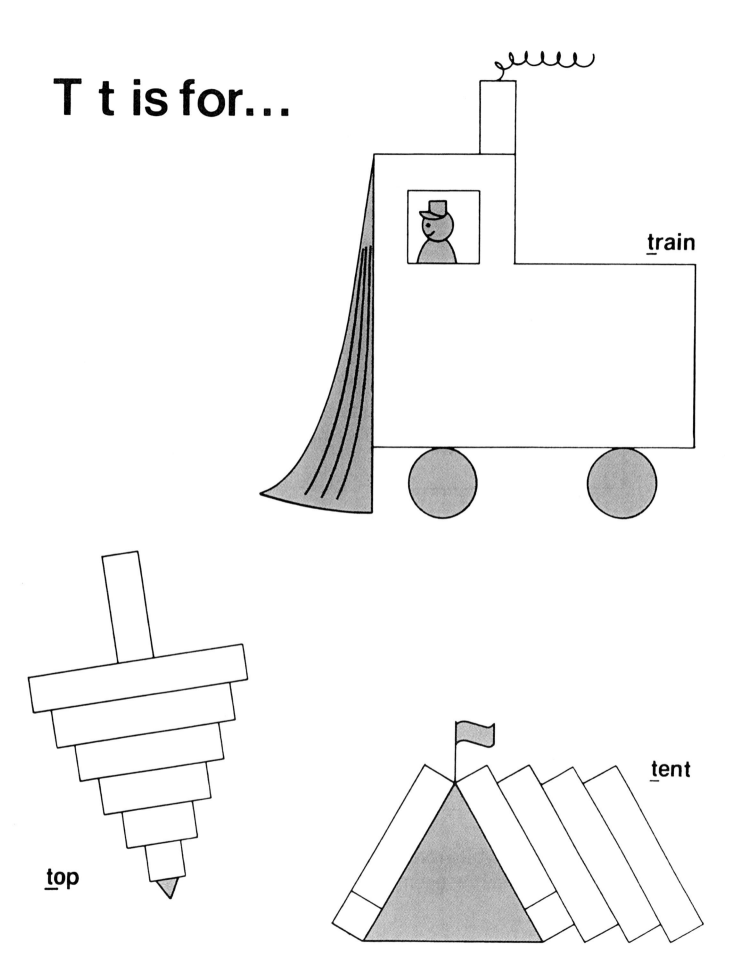

train

top

tent

1. Build the U with rods. What colors did you use?

   _____

   How many did you use?

   _____

2. Use exactly 6 rods to build the U.

   What colors did you use?

   _____

3. Build the U using rods of all different colors.

   Color your U.

4. Cover each color with white rods.

   How many white rods did you need for each color?

   _____

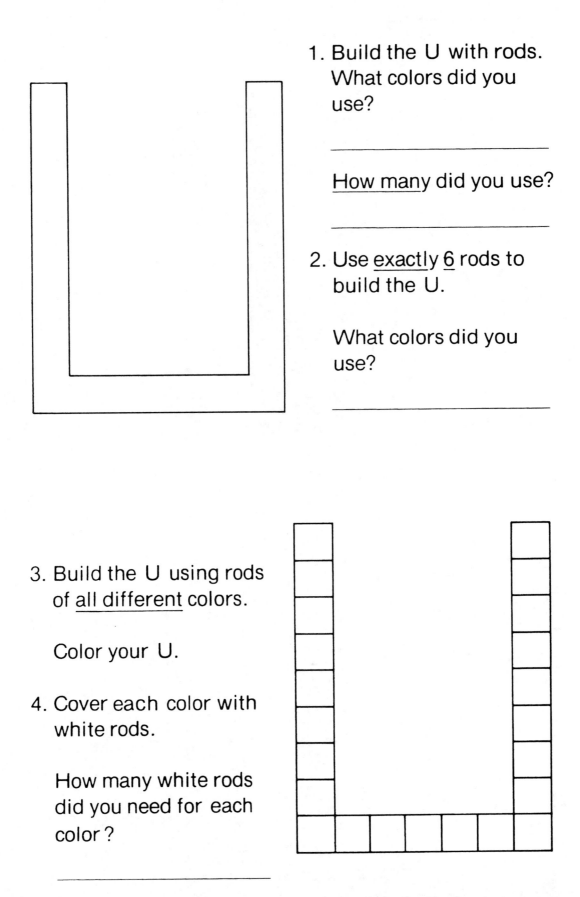

Cuisenaire® Rods Alphabet Book   © Learning Resources, Inc.

# U u is for...

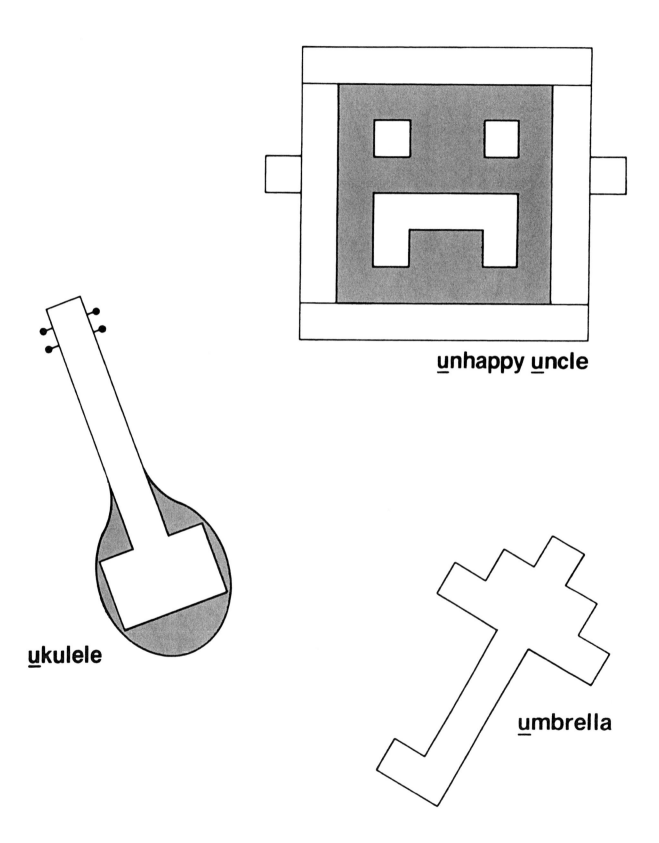

**<u>u</u>nhappy <u>u</u>ncle**

**<u>u</u>kulele**

**<u>u</u>mbrella**

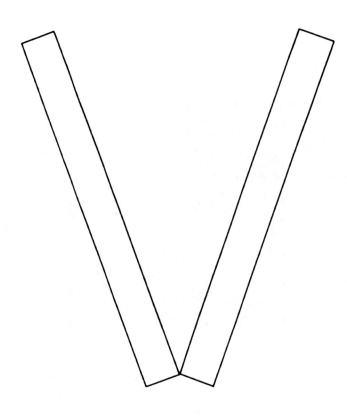

1. Build the V with rods. What colors did you use?

   _____

   How many rods did you use?

   _____

2. Use 4 rods of the same color to build the V.

   What rod did you use?

   _____

3. Build the V using rods of all different colors.

   Color your V.

4. Cover each color with white rods.

   How many white rods did you need for each color?

   _____

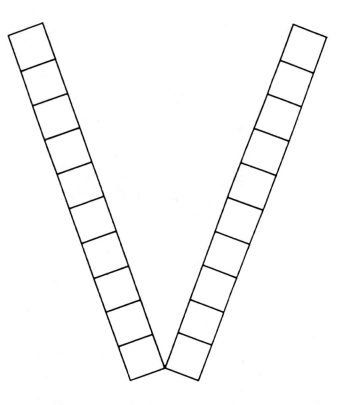

Cuisenaire® Rods Alphabet Book   © Learning Resources, Inc.

# V v is for...

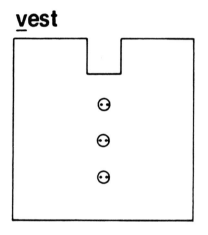

vest

van

valentine

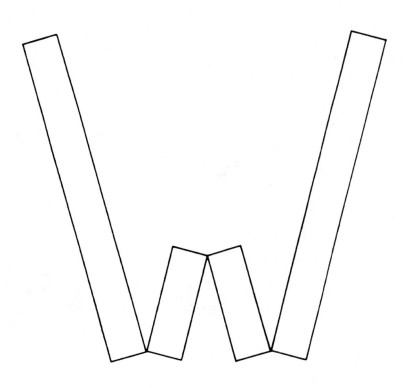

1. Build the W with rods. What colors did you use?

   _____

   How many rods did you use?

   _____

2. Use <u>more than 6</u> rods to build the W.

   What colors did you use?

   _____

3. Build the W with rods of <u>all different</u> colors.

   Color your W.

4. Cover each color with white rods.

   How many white rods did you need for each color?

   _____

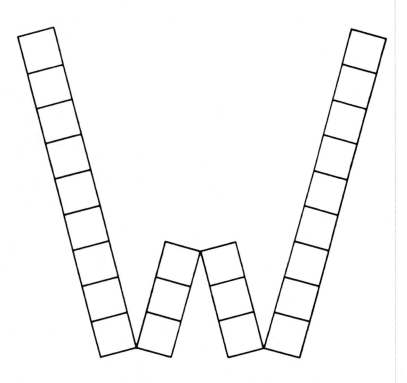

Cuisenaire® Rods Alphabet Book   © Learning Resources, Inc.

# W w is for...

**w**eb

**w**indow

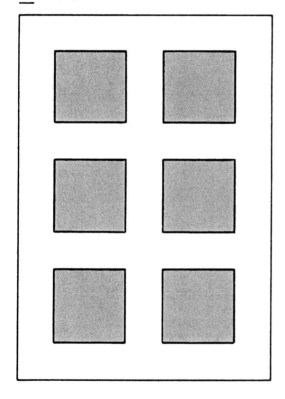

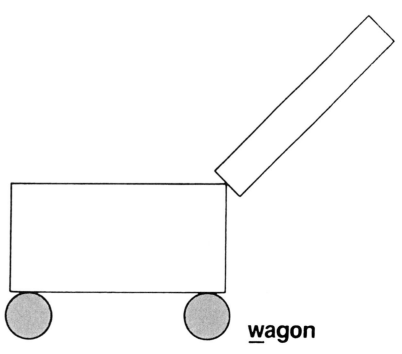

**w**agon

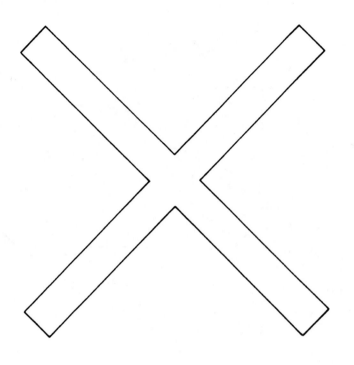

1. Build an X with rods. What colors did you use?

   _____

   How many rods did you use?

   _____

2. Use <u>less than 5</u> rods to build the X.

   What colors did you use?

   _____

3. Build the X with rods of <u>all different</u> colors.

   Color your X.

4. Cover each color with white rods.

   How many white rods did you need for each color?

   _____

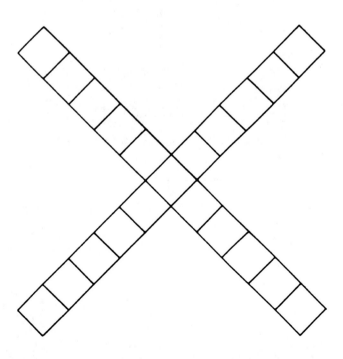

# X x is for...

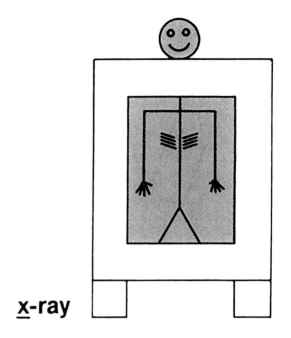

**x-ray**

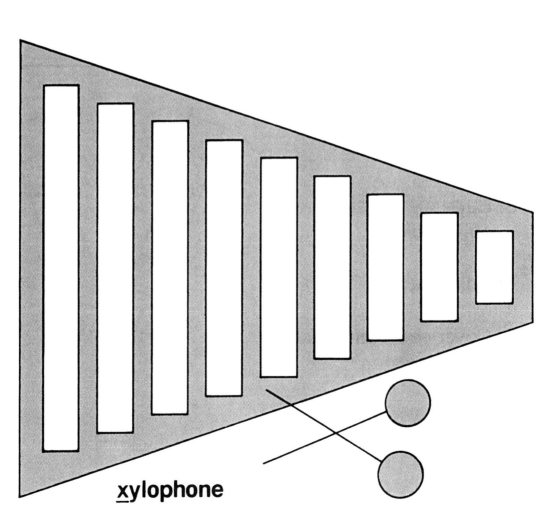

**xylophone**

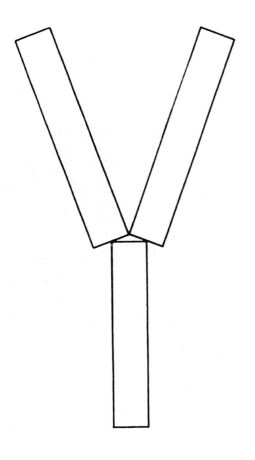

1. Build the Y with rods. What colors did you use?

   _____

   How many rods did you use?

   _____

2. Use <u>more than 8</u> rods to build the Y.

   What colors did you use?

   _____

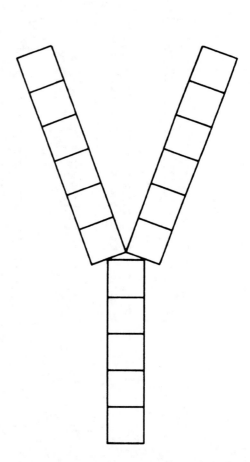

3. Build the Y using <u>only 2</u> colors.

   Color your Y.

4. Cover each color with white rods.

   How many white rods did you need for each color?

   _____

Cuisenaire® Rods Alphabet Book   © Learning Resources, Inc.

# Y y is for...

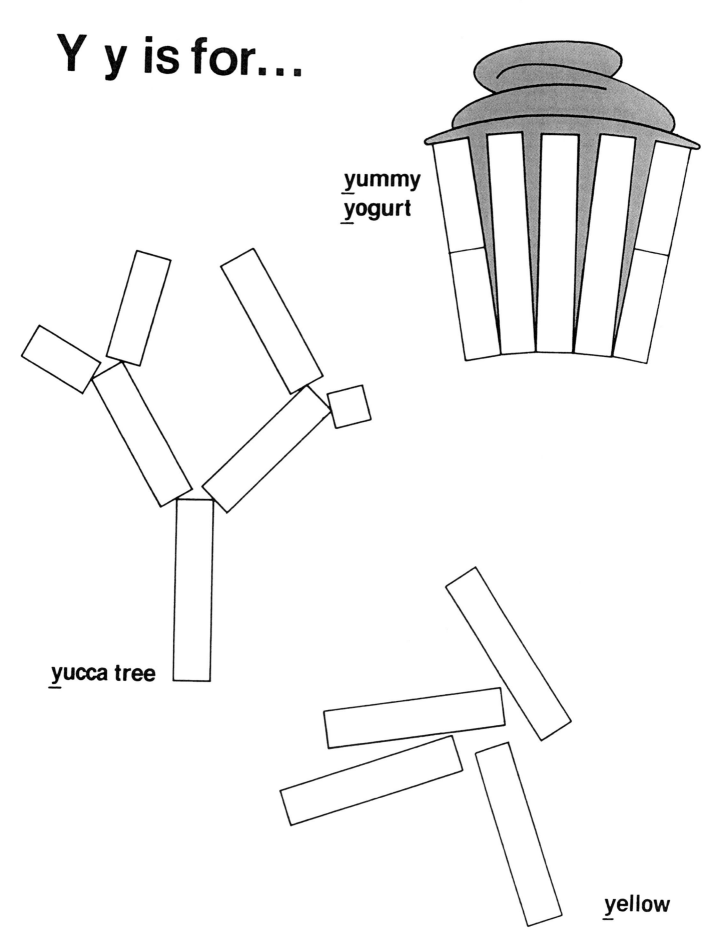

<u>y</u>ummy
<u>y</u>ogurt

<u>y</u>ucca tree

<u>y</u>ellow

Cuisenaire® Rods Alphabet Book   © Learning Resources, Inc.

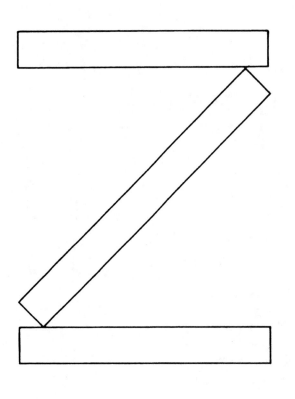

1. Build the Z with rods. What colors did you use?

   _____

   How many rods did you use?

   _____

2. Use <u>exactly</u> <u>15</u> rods to build the Z.

   What colors did you use?

   _____

3. Build the Z with rods of all <u>different</u> colors.

   Color your Z.

4. Cover each color with white rods.

   How many white rods did you need for each color?

   _____

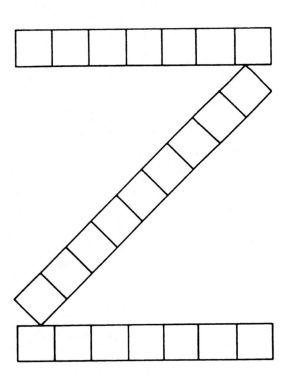

Cuisenaire® Rods Alphabet Book   © Learning Resources, Inc.

# Z z is for...

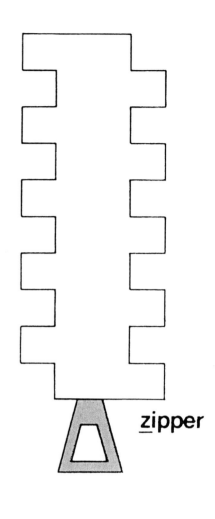
zipper

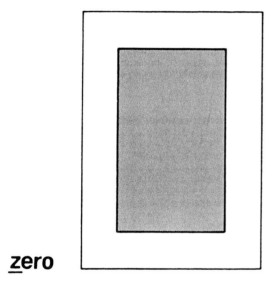
zero

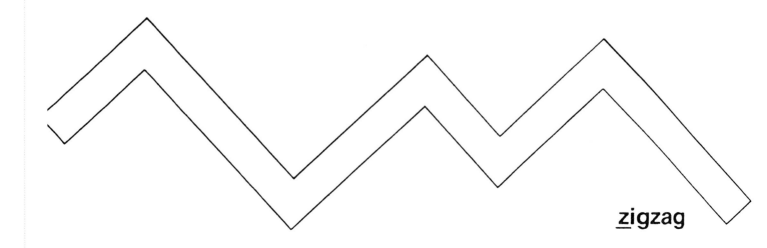
zigzag

# ANSWERS AND SUGGESTIONS

## A

**Mathematics Vocabulary and Concepts:**

**how many** (counting colors and rods), **only light green** (pre-multiplication, facts for 3), **all different** (problem solving, no two rods of the same color), **cover** (matching white rods to centimeter squares, one-to-one correspondence), **how many white rods** (counting and sums for 21).

### Addition Stories:

| Some Rod Stories | Number Stories When White = 1 |
|---|---|
| E + G + E | 9 + 3 + 9 = 21 |
| G + G + G + G + G + G + G | 3 + 3 + 3 + 3 + 3 + 3 + 3 = 21 |
| | or 7 × 3 = 21 |
| W + N + D + G + G | 1 + 8 + 6 + 3 + 3 = 21 |
| R + K + W + N + G | 2 + 7 + 1 + 8 + 3 = 21 |
| P + Y + G + W + D + R | 4 + 5 + 3 + 1 + 6 + 2 = 21 |

**Other A Words:** arch, ax, antler, arrow

## B

**Mathematics Vocabulary and Concepts:**

**how many** (counting colors and rods), **exactly 9 rods** (counting, problem solving, more than one right answer), **only 2 colors** (problem solving, more than one right answer), **cover** (matching white rods to centimeter squares, one-to-one correspondence), **how many white rods** (counting and sums for 27).

### Addition Stories:

| Some Rod Stories | Number Stories When White = 1 |
|---|---|
| E + G + G + W + G + R + R + R + R | 9 + 3 + 3 + 1 + 3 + 2 + 2 + 2 + 2 = 27 |
| G + G + G + G + G + R + R + R + R + R + R | 3 + 3 + 3 + 3 + 3 + 2 + 2 + 2 + 2 + 2 + 2 = 27 |
| | or (5 × 3) + (6 × 2) = 27 |
| P + P + P + P + G + R + R + R + R | 4 + 4 + 4 + 4 + 3 + 2 + 2 + 2 + 2 = 27 |
| | or (4 × 4) + (1 × 3) + (4 × 2) = 27 |
| P + P + G + W + W + R + R + R + R + R + R + R | 4 + 4 + 3 + 1 + 1 + 2 + 2 + 2 + 2 + 2 + 2 + 2 = 27 |
| | or (2 × 4) + (1 × 3) + (2 × 1) + (7 × 2) = 27 |

**Other B Words:** big, bird, bag

## C

**Mathematics Vocabulary and Concepts:**

**how many** (counting colors and rods), **the least number** (minimum, maximum, problem solving, longest rods), **all different** (problem solving, no two rods of the same color), **cover** (matching white rods to centimeter squares, one-to-one correspondence), **how many white rods** (counting and sums for 22).

### Addition Stories:

| Some Rod Stories | Number Stories When White = 1 |
|---|---|
| E + D + K | 9 + 6 + 7 = 22 |
| N + K + K | 8 + 7 + 7 = 22 |
| E + P + R + K | 9 + 4 + 2 + 7 = 22 |
| O + W + Y + D | 10 + 1 + 5 + 6 = 22 |
| K + G + R + P + D | 7 + 3 + 2 + 4 + 6 = 22 |

**Other C Words:** cart, cat, camera, clock

## D

**Mathematics Vocabulary and Concepts:**

**how many** (counting colors and rods), **more than 6** (greater than, inequalities, problem solving, more than one right answer) **only 2 colors** (problem solving, more than one right answer), **cover** (matching white rods to centimeter squares, one-to-one correspondence), **how many white rods** (counting and sums for 26).

### Addition Stories:

| Some Rod Stories | Number Stories When White = 1 |
|---|---|
| G + G + G + G + R + R + G + R + R + G | 3 + 3 + 3 + 3 + 2 + 2 + 3 + 2 + 2 + 3 = 26 |
| E + Y + Y + R + G + R | 9 + 5 + 5 + 2 + 3 + 2 = 26 |
| E + P + W + P + W + R + G + R | 9 + 4 + 1 + 4 + 1 + 2 + 3 + 2 = 26 |
| D + G + Y + Y + R + G + R | 6 + 3 + 5 + 5 + 2 + 3 + 2 = 26 |
| Y + G + Y + W + Y + W + W + G + R | 5 + 3 + 5 + 1 + 5 + 1 + 1 + 3 + 2 = 26 |

**Other D Words:** doll, duck, donut, dump truck

### E

**Mathematics Vocabulary and Concepts:**

**how many** (counting colors and rods), **only red** (pre-multiplication, facts for 2), **only 2 colors** (problem solving, more than one right answer), **cover** (matching white rods to centimeter squares, one-to-one correspondence), **how many white rods** (counting and sums for 20).

#### Addition Stories:

| Some Rod Stories | Number Stories When White = 1 |
|---|---|
| R + R + R + R + R + R + R + R + R + R | 2 + 2 + 2 + 2 + 2 + 2 + 2 + 2 + 2 + 2 = 20 |
| | or 10 × 2 = 20 |
| R + R + R + R + R + R + R + R + W + W + W + W | 2 + 2 + 2 + 2 + 2 + 2 + 2 + 2 + 1 + 1 + 1 + 1 = 20 |
| P + P + P + P + R + R | 4 + 4 + 4 + 4 + 2 + 2 = 20 |
| D + R + R + R + R + R + R + R | 6 + 2 + 2 + 2 + 2 + 2 + 2 + 2 = 20 |
| N + P + P + R + R | 8 + 4 + 4 + 2 + 2 = 20 |

**Other E Words:** Easter egg, eight, elf, engine

### F

**Mathematics Vocabulary and Concepts:**

**how many** (counting colors and rods), **exactly 5 rods** (counting, problem solving, more than one right answer), **all different** (problem solving, no two rods of the same color), **cover** (matching white rods to centimeter squares, one-to-one correspondence), **how many white rods** (counting and sums for 14).

#### Addition Stories:

| Some Rod Stories | Number Stories When White = 1 |
|---|---|
| P + G + G + R + R | 4 + 3 + 3 + 2 + 2 = 14 |
| P + D + R + W + W | 4 + 6 + 2 + 1 + 1 = 14 |
| G + G + R + R + R + R | 3 + 3 + 2 + 2 + 2 + 2 = 14 |
| P + Y + G + R | 4 + 5 + 3 + 2 = 14 |
| W + P + K + R | 1 + 4 + 7 + 2 = 14 |

**Other F Words:** five, faucet, fox, fan

### G

**Mathematics Vocabulary and Concepts:**

**how many** (counting colors and rods), **only light green** (pre-multiplication, facts for 3), **all different** (problem solving, no two rods of the same color), **cover** (matching white rods to centimeter squares, one-to-one correspondence), **how many white rods** (counting and sums for 27).

#### Addition Stories:

| Some Rod Stories | Number Stories When White = 1 |
|---|---|
| G + G + G + G + G + G + G + G + G | 3 + 3 + 3 + 3 + 3 + 3 + 3 + 3 + 3 = 27 |
| | or 9 × 3 = 27 |
| D + O + Y + G + W + R | 6 + 10 + 5 + 3 + 1 + 2 = 27 |
| D + E + Y + P + W + R | 6 + 9 + 5 + 4 + 1 + 2 = 27 |
| D + N + R + P + P + G | 6 + 8 + 2 + 4 + 4 + 3 = 27 |
| Y + K + P + Y + G + G | 5 + 7 + 4 + 5 + 3 + 3 = 27 |

**Other G Words:** garage, grasshopper, gun, gate

### H

**Mathematics Vocabulary and Concepts:**

**how many** (counting colors and rods), **less than 6** (inequalities, problem solving, more than one right answer), **all different** (problem solving, no two rods of the same color), **cover** (matching white rods to centimeter squares, one-to-one correspondence), **how many white rods** (counting and sums for 22).

#### Addition Stories:

| Some Rod Stories | Number Stories When White = 1 |
|---|---|
| E + E + P | 9 + 9 + 4 = 22 |
| P + P + Y + E | 4 + 4 + 5 + 9 = 22 |
| R + K + P + E | 2 + 7 + 4 + 9 = 22 |
| W + N + P + E | 1 + 8 + 4 + 9 = 22 |
| P + Y + W + G + E | 4 + 5 + 1 + 3 + 9 = 22 |

**Other H Words:** horse, happy, Halloween pumpkin, hamburger

## I

**Mathematics Vocabulary and Concepts:**

**how many** (counting colors and rods), **only one color** (problem solving, only one right answer), **all different** (problem solving, no two rods of the same color), **cover** (matching white rods to centimeter squares, one-to-one correspondence), **how many white rods** (counting and sums for 16).

### Addition Stories:

| **Some Rod Stories** | **Number Stories When White = 1** |
|---|---|
| Y + D + W + P | 5 + 6 + 1 + 4 = 16 |
| G + R + D + W + P | 3 + 2 + 6 + 1 + 4 = 16 |
| Y + Y + W + Y | 5 + 5 + 1 + 5 = 16 |
| W + W + W + W + W + W + W + W + W + W + W + W + W + W + W + W | 1 + 1 + 1 + 1 + 1 + 1 + 1 + 1 + 1 + 1 + 1 + 1 + 1 + 1 + 1 + 1 = 16 |
| R + R + N + R + R | 2 + 2 + 8 + 2 + 2 = 16 |

**Other I Words:** instrument, inside, icicle, ice cream cone

## J

**Mathematics Vocabulary and Concepts:**

**how many** (counting colors and rods), **exactly 6 rods** (counting, problem solving, more than one right answer), **all different** (problem solving, no two rods of the same color), **cover** (matching white rods to centimeter squares, one-to-one correspondence), **how many white rods** (counting and sums for 17).

### Addition Stories:

| **Some Rod Stories** | **Number Stories When White = 1** |
|---|---|
| Y + P + R + W + G + R | 5 + 4 + 2 + 1 + 3 + 2 = 17 |
| Y + K + R + W + W + W | 5 + 7 + 2 + 1 + 1 + 1 = 17 |
| Y + K + G + R | 5 + 7 + 3 + 2 = 17 |
| Y + P + G + G + R | 5 + 4 + 3 + 3 + 2 = 17 |
| Y + W + D + G + R | 5 + 1 + 6 + 3 + 2 = 17 |

**Other J Words:** jet, jeep, jail, jelly jar

## K

**Mathematics Vocabulary and Concepts:**

**how many** (counting colors and rods), **only yellow** (pre-multiplication, facts for 5), **all different** (problem solving, no two rods of the same color), **cover** (matching white rods to centimeter squares, one-to-one correspondence), **how many white rods** (counting and sums for 20).

### Addition Stories:

| **Some Rod Stories** | **Number Stories When White = 1** |
|---|---|
| Y + Y + Y + Y | 5 + 5 + 5 + 5 = 20 <br> or 4 × 5 = 20 |
| O + P + W + R + G | 10 + 4 + 1 + 2 + 3 = 20 |
| N + R + P + W + Y | 8 + 2 + 4 + 1 + 5 = 20 |
| E + W + G + R + Y | 9 + 1 + 3 + 2 + 5 = 20 |
| P + D + Y + G + R | 4 + 6 + 5 + 3 + 2 = 20 |

**Other K Words:** kitten, key, kingdom, kettle

## L

**Mathematics Vocabulary and Concepts:**

**how many** (counting colors and rods), **exactly 5 rods** (counting, problem solving, more than one right answer), **all different** (problem solving, no two rods of the same color), **cover** (matching white rods to centimeter squares, one-to-one correspondence), **how many white rods** (counting and sums for 16).

### Addition Stories:

| **Some Rod Stories** | **Number Stories When White = 1** |
|---|---|
| D + R + W + Y + R | 6 + 2 + 1 + 5 + 2 = 16 |
| Y + G + R + W + Y | 5 + 3 + 2 + 1 + 5 = 16 |
| G + K + P + R | 3 + 7 + 4 + 2 = 16 |
| R + N + G + G | 2 + 8 + 3 + 3 = 16 |
| E + W + D | 9 + 1 + 6 = 16 |

**Other L Words:** lady, little, ladybug, lion

## M

**Mathematics Vocabulary and Concepts:**

**how many** (counting colors and rods), **the least number** (minimum, maximum, problem solving, longest rods), **all different** (problem solving, no two rods of the same color, only one right answer), **cover** (matching white rods to centimeter squares, one-to-one correspondence), **how many white rods** (counting and sums for 26).

### Addition Stories:

| Some Rod Stories | Number Stories When White = 1 |
|---|---|
| E + P + E + P | 9 + 4 + 9 + 4 = 26 |
| R + K + P + G + W + E | 2 + 7 + 4 + 3 + 1 + 9 = 26 |
| Y + P + G + W + P + D + G | 5 + 4 + 3 + 1 + 4 + 6 + 3 = 26 |
| K + R + P + R + R + Y + P | 7 + 2 + 4 + 2 + 2 + 5 + 4 = 26 |
| D + G + G + W + P + E | 6 + 3 + 3 + 1 + 4 + 9 = 26 |

**Other M Words:** magnet, motorcycle, microphone, Mother

## N

**Mathematics Vocabulary and Concepts:**

**how many** (counting colors and rods), **more than 7** (greater than, inequalities, problem solving, more than one right answer), **all different** (problem solving, no two rods of the same color), **cover** (matching white rods to centimeter squares, one-to-one correspondence), **how many white rods** (counting and sums for 29).

### Addition Stories:

| Some Rod Stories | Number Stories When White = 1 |
|---|---|
| K + G + P + Y + W + R + D + W | 7 + 3 + 4 + 5 + 1 + 2 + 6 + 1 = 29 |
| P + P + R + G + G + G + P + P + R | 4 + 4 + 2 + 3 + 3 + 3 + 4 + 4 + 2 = 29 |
| W + E + R + K + O' | 1 + 9 + 2 + 7 + 10 = 29 |
| W + E + R + K + P + D | 1 + 9 + 2 + 7 + 4 + 6 = 29 |
| R + N + E + G + K | 2 + 8 + 9 + 3 + 7 = 29 |

**Other N Words:** net, needle, nose, noose

## O

**Mathematics Vocabulary and Concepts:**

**how many** (counting colors and rods), **exactly 10 rods** (counting, problem solving, more than one right answer), **only 2 colors** (problem solving, more than one right answer), **cover** (matching white rods to centimeter squares, one-to-one correspondence), **how many white rods** (counting and sums for 24).

### Addition Stories:

| Some Rod Stories | Number Stories When White = 1 |
|---|---|
| P + R + G + W + R + P + R + W + G + R | 4 + 2 + 3 + 1 + 2 + 4 + 2 + 1 + 3 + 2 = 24 |
| P + W + W + P + R + P + R + P + W + W | 4 + 1 + 1 + 4 + 2 + 4 + 2 + 4 + 1 + 1 = 24 |
| P + R + P + R + P + R + P + R | 4 + 2 + 4 + 2 + 4 + 2 + 4 + 2 = 24 |
| | or (4 × 4) + (4 × 2) = 24 |
| P + W + W + P + W + W + P + W + W + P + W + W | 4 + 1 + 1 + 4 + 1 + 1 + 4 + 1 + 1 + 4 + 1 + 1 = 24 |
| | or (4 × 4) + (8 × 1) = 24 |
| G + W + R + G + W + R + G + W + R + G + W + R | 3 + 1 + 2 + 3 + 1 + 2 + 3 + 1 + 2 + 3 + 1 + 2 = 24 |
| | or (4 × 3) + (4 × 1) + (4 × 2) = 24 |

**Other O Words:** owl, oval, out, octagon

## P

**Mathematics Vocabulary and Concepts:**

**how many** (counting colors and rods), **only three colors** (problem solving, more than one right answer), **all different** (problem solving, no two rods of the same color). **cover** (matching white rods to centimeter squares, one-to-one correspondence), **how many white rods** (counting and sums for 25).

### Addition Stories:

| Some Rod Stories | Number Stories When White = 1 |
|---|---|
| Y + Y + Y + Y + R + G | 5 + 5 + 5 + 5 + 2 + 3 = 25 |
| | or (4 × 5) + 2 + 3 = 25 |
| D + P + Y + Y + Y | 6 + 4 + 5 + 5 + 5 = 25 |
| W + O' + Y + R + G + P | 1 + 10 + 5 + 2 + 3 + 4 = 25 |
| R + N + D + Y + P | 2 + 8 + 6 + 5 + 4 = 25 |
| G + K + D + Y + P | 3 + 7 + 6 + 5 + 4 = 25 |

**Other P Words:** parrot, pole, patch, pentagon

## Q

**Mathematics Vocabulary and Concepts:**

**how many** (counting colors and rods), **the least number** (minimum, maximum, problem solving, longest rods), **all different** (problem solving, no two rods of the same color), **cover** (matching white rods to centimeter squares), **how many white rods** (counting and sums for 30).

**Addition Stories:**

| Some Rod Stories | Number Stories When White = 1 |
|---|---|
| N + K + K + D + R | 8 + 7 + 7 + 6 + 2 = 30 |
| E + Y + E + Y + R | 9 + 5 + 9 + 5 + 2 = 30 |
| E + Y + N + D + R | 9 + 5 + 8 + 6 + 2 = 30 |
| N + D + K + G + P + R | 8 + 6 + 7 + 3 + 4 + 2 = 30 |
| E + P + W + Y + G + D + R | 9 + 4 + 1 + 5 + 3 + 6 + 2 = 30 |

**Other Q Words:** quacking duck, quotation marks, quail, quiet

## R

**Mathematics Vocabulary and Concepts:**

**how many** (counting colors and rods), **exactly 8 rods** (counting, problem solving, more than one right answer), **only 2 colors** (problem solving, more than one right answer), **cover** (matching white rods to centimeter squares, one-to-one correspondence), **how many white rods** (counting and sums for 34).

**Addition Stories:**

| Some Rod Stories | Number Stories When White = 1 |
|---|---|
| R + R + R + G + G + Y + Y + P + P + P | 2 + 2 + 2 + 3 + 3 + 5 + 5 + 4 + 4 + 4 = 34 |
| | or $(3 \times 2) + (2 \times 3) + (2 \times 5) + (3 \times 4) = 34$ |
| D + D + D + R + R + R + R + R + R + R + R | 6 + 6 + 6 + 2 + 2 + 2 + 2 + 2 + 2 + 2 + 2 = 34 |
| | or $(3 \times 6) + (8 \times 2) = 34$ |
| Y + Y + Y + K + D + R + R + R | 5 + 5 + 5 + 7 + 6 + 2 + 2 + 2 = 34 |
| P + P + P + P + P + R + R + R + R + R + R + R | 4 + 4 + 4 + 4 + 4 + 2 + 2 + 2 + 2 + 2 + 2 + 2 = 34 |
| | or $(5 \times 4) + (7 \times 2) = 34$ |

**Other R Words:** robot, rattle, rug, radio

## S

**Mathematics Vocabulary and Concepts:**

**how many** (counting colors and rods), **only light green** (pre-multiplication, facts for 3), **only 2 colors** (problem solving, more than one right answer), **cover** (matching white rods to centimeter squares, one-to-one correspondence), **how many white rods** (counting and sums for 30).

**Addition Stories:**

| Some Rod Stories | Number Stories When White = 1 |
|---|---|
| G + G + G + G + G + G + G + G + G + G | 3 + 3 + 3 + 3 + 3 + 3 + 3 + 3 + 3 + 3 = 30 |
| | or $10 \times 3 = 30$ |
| D + D + D + G + G + G + G | 6 + 6 + 6 + 3 + 3 + 3 + 3 = 30 |
| | or $(3 \times 6) + (4 \times 3) = 30$ |
| G + R + R + G + R + G + G + G + G + R + R + R | 3 + 2 + 2 + 3 + 2 + 3 + 3 + 3 + 3 + 2 + 2 + 2 = 30 |
| | or $(6 \times 3) + (6 \times 2) = 30$ |
| R + P + R + P + R + P + R + P + R + P | 2 + 4 + 2 + 4 + 2 + 4 + 2 + 4 + 2 + 4 = 30 |
| | or $(5 \times 2) + (5 \times 4) = 30$ |

**Other S Words:** staircase, star, submarine, shovel

## T

**Mathematics Vocabulary and Concepts:**

**how many** (counting colors and rods), **only one color** (problem solving, pre-multiplication, facts for 3), **all different** (problem solving, no two rods of the same color), **cover** (matching white rods to centimeter squares, one-to-one correspondence), **how many white rods** (counting and sums for 15).

**Addition Stories:**

| Some Rod Stories | Number Stories When White = 1 |
|---|---|
| G + G + G + G + G | 3 + 3 + 3 + 3 + 3 = 15 |
| | or $5 \times 3 = 15$ |
| K + N | 7 + 8 = 15 |
| G + P + R + D | 3 + 4 + 2 + 6 = 15 |
| W + D + G + Y | 1 + 6 + 3 + 5 = 15 |
| R + Y + W + G + P | 2 + 5 + 1 + 3 + 4 = 15 |

**Other T Words:** table, truck, thermometer, taxi

## U

**Mathematics Vocabulary and Concepts:**

**how many** (counting colors and rods), **exactly 6 rods** (counting, problem solving, more than one right answer), **all different** (problem solving, no two rods of the same color), **cover** (matching white rods to centimeter squares, one-to-one correspondence), **how many white rods** (counting and sums to 23).

### Addition Stories:

| Some Rod Stories | Number Stories When White = 1 |
|---|---|
| R + D + W + D + G + Y | 2 + 6 + 1 + 6 + 3 + 5 = 23 |
| W + R + Y + K + P + P | 1 + 2 + 5 + 7 + 4 + 4 = 23 |
| E + D + N | 9 + 6 + 8 = 23 |
| K + W + D + E | 7 + 1 + 6 + 9 = 23 |
| N + R + Y + W + K | 8 + 2 + 5 + 1 + 7 = 23 |

**Other U Words:** unicorn, undershirt, ugly monster, upside-down

## V

**Mathematics Vocabulary and Concepts:**

**how many** (counting colors and rods), **4 rods of the same color** (problem solving, pre-multiplication, facts for 5), **all different** (problem solving, no two rods of the same color), **cover** (matching white rods to centimeter squares, one-to-one correspondence), **how many white rods** (counting and sums to 20).

### Addition Stories:

| Some Rod Stories | Number Stories When White = 1 |
|---|---|
| Y + Y + Y + Y | 5 + 5 + 5 + 5 = 20 |
| O + R + G + Y | 10 + 2 + 3 + 5 = 20 |
| W + E + R + N | 1 + 9 + 2 + 8 = 20 |
| W + E + R + G + Y | 1 + 9 + 2 + 3 + 5 = 20 |
| R + N + W + G + D | 2 + 8 + 1 + 3 + 6 = 20 |

**Other V Words:** violin, violet, vegetable, vampire

## W

**Mathematics Vocabulary and Concepts:**

**how many** (counting colors and rods), **more than 6** (greater than, inequalities, problem solving, more than one right answer), **all different** (problem solving, no two rods of the same color), **cover** (matching white rods to centimeter squares, one-to-one correspondence), **how many white rods** (counting and sums for 24).

### Addition Stories:

| Some Rod Stories | Number Stories When White = 1 |
|---|---|
| D + G + R + W + W + R + G + D | 6 + 3 + 2 + 1 + 1 + 2 + 3 + 6 = 24 |
| G + G + G + G + G + G + G + G | 3 + 3 + 3 + 3 + 3 + 3 + 3 + 3 = 24 |
| | or 8 × 3 = 24 |
| P + P + W + G + G + W + P + P | 4 + 4 + 1 + 3 + 3 + 1 + 4 + 4 = 24 |
| | or (4 × 4) + (2 × 1) + (2 × 3) = 24 |
| Y + P + G + G + P + Y | 5 + 4 + 3 + 3 + 4 + 5 = 24 |
| | or (2 × 5) + (2 × 4) + (2 × 3) = 24 |
| E + G + W + R + P + Y | 9 + 3 + 1 + 2 + 4 + 5 = 24 |

**Other W Words:** whale, windmill, wizard, whistle

## X

**Mathematics Vocabulary and Concepts:**

**how many** (counting colors and rods), **less than 5** (less than, inequalities, problem solving, more than one right answer for 4 rods, but none for 1, 2 or 3), **all different** (problem solving, no two rods of the same color), **cover** (matching white rods to centimeter squares, one-to-one correspondence), **how many white rods** (counting and sums for 21).

### Addition Stories:

| Some Rod Stories | Number Stories When White = 1 |
|---|---|
| Y + Y + Y + D | 5 + 5 + 5 + 6 = 21 |
| Y + Y + O + W | 5 + 5 + 10 + 1 = 21 |
| Y + D + G + R + P + W | 5 + 6 + 3 + 2 + 4 + 1 = 21 |
| Y + P + W + E + R | 5 + 4 + 1 + 9 + 2 = 21 |
| Y + P + W + N + G | 5 + 4 + 1 + 8 + 3 = 21 |

**Other X Words:** Since words beginning with X are so sparse, words which contain an X might be used: fox, exit, box, ox

## Y Mathematics Vocabulary and Concepts:

**how many** (counting colors and rods), **more than 8** (greater than, inequalities, problem solving, more than one right answer), **only 2 colors** (problem solving, more than one right answer), **cover** (matching white rods to centimeter squares, one-to-one correspondence), **how many white rods** (counting and sums for 17).

### Addition Stories:

**Some Rod Stories**
W + R + G + W + R + G + W + R + R
R + R + R + R + R + R + R + R + W

R + R + R + R + R + R + Y

D + D + Y
G + G + G + G + Y

**Number Stories When White = 1**
1 + 2 + 3 + 1 + 2 + 3 + 1 + 2 + 2 = 17
2 + 2 + 2 + 2 + 2 + 2 + 2 + 2 + 1 = 17
or (8 × 2) + 1 = 17
2 + 2 + 2 + 2 + 2 + 5 = 17
or (6 × 2) + 5 = 17
6 + 6 + 5 = 17   or (6 × 2) + 5 = 17
3 + 3 + 3 + 3 + 5 = 17
or (4 × 3) + 5 = 17

**Other Y Words:** yacht, yolk, yoyo, youth

## Z Mathematics Vocabulary and Concepts:

**how many** (counting colors and rods), **exactly 15 rods** (counting, problem solving, more than one right answer), **all different** (problem solving, no two rods of the same color), **cover** (matching white rods to centimeter squares, one-to-one correspondence), **how many white rods** (counting and sums for 23).

### Addition Stories:

**Some Rod Stories**
W + W + W + W + W + W + W + W + W +
W + W + W + W + W + W + K
W + W + W + W + W + W + W + G + G + G +
W + W + W + W + W + W + W

W + D + R + K + G + P
E + W + D + R + Y
G + D + K + W + R + P

**Number Stories When White = 1**
1 + 1 + 1 + 1 + 1 + 1 + 1 + 1 + 1 + 1 + 1 + 1 + 1 +
1 + 1 + 1 + 7 = 23
1 + 1 + 1 + 1 + 1 + 1 + 1 + 3 + 3 + 3 + 1 + 1 + 1 +
1 + 1 + 1 + 1 = 23
or (14 × 1) + (3 × 3) = 23
1 + 6 + 2 + 7 + 3 + 4 = 23
9 + 1 + 6 + 2 + 5 = 23
3 + 6 + 7 + 1 + 2 + 4 = 23

**Other Z Words:** zebra, zoom, zap, zoo

# GLOSSARY OF ALPHABET WORDS USED IN THIS BOOK

airplane
alligator
apple tree

bed
boat
building

candle
clown
cup

dog
dragon

ears
elephant

fish tank
flag
flower

gift
grass
grin

hammer
hat
house

igloo
Indian
insect

jacks
jug

kick
king
kite

lamp
lemon lollipop
log cabin

mail box
man
milk bottle

nails
nest

nutcracker
nuts

octopus
ostrich

pail
paintbrush
pals

queen
question mark
quilt

rake
rocket
ruler

seven
snake
sun

tent
top
train

ukelele
umbrella
unhappy uncle

valentine
van
vest

wagon
web
window

x-ray
xylophone

yellow
yucca tree
yummy yogurt

zero
zigzag
zipper